PREALGEBRA SKILLS
WORKBOOK

JOSHUA YARMISH
Pace University

RACHEL STURM-BEISS
Kingsborough Community College

PEARSON
Prentice
Hall

Upper Saddle River, NJ 07458

Editor-in-Chief: Chris Hoag
Senior Acquisitions Editor: Paul Murphy
Supplement Editor: Christina Simoneau
Executive Managing Editor: Kathleen Schiaparelli
Assistant Managing Editor: Becca Richter
Production Editor: Donna Crilly
Supplement Cover Manager: Paul Gourhan
Supplement Cover Designer: Joanne Alexandris
Manufacturing Buyer: Michael Bell

© 2005 Pearson Education, Inc.
Pearson Prentice Hall
Pearson Education, Inc.
Upper Saddle River, NJ 07458

Pearson Prentice Hall™ is a trademark of Pearson Education, Inc.

The author and publisher of this book have used their best efforts in preparing this book. These efforts include the development, research, and testing of the theories and programs to determine their effectiveness. The author and publisher make no warranty of any kind, expressed or implied, with regard to these programs or the documentation contained in this book. The author and publisher shall not be liable in any event for incidental or consequential damages in connection with, or arising out of, the furnishing, performance, or use of these programs.

Printed in the United States of America

10 9 8 7 6 5 4 3 2 1

ISBN 0-13-186920-5

Pearson Education Ltd., *London*
Pearson Education Australia Pty. Ltd., *Sydney*
Pearson Education Singapore, Pte. Ltd.
Pearson Education North Asia Ltd., *Hong Kong*
Pearson Education Canada, Inc., *Toronto*
Pearson Educación de Mexico, S.A. de C.V.
Pearson Education—Japan, *Tokyo*
Pearson Education Malaysia, Pte. Ltd.

PREALGEBRA SKILLS WORKBOOK

Table of Contents

Chapter 5: Ratios and Proportions

Chapter 6: Percentages

Chapter 7: Averages

Solutions

PREFACE

This workbook is designed as a resource to help students acquire a solid foundation in prealgebra through the review and repetition of basic topics. By doing so, this workbook is also intended to help prepare students for success in their future algebra/mathematics courses. All topics and concepts in the workbook are presented and organized in a format that is easy for students to follow and convenient for instructors to use as a resource for lesson planning and problem selection. The sections are full of worked-out examples that illustrate techniques that can be applied to end-of-section exercises. The end-of-section exercises were carefully formulated to reinforce basic skills and to build upon these skills with the goal of acquiring the techniques and methods for solving more complex problems. Throughout the workbook there is an emphasis on mathematical reasoning and problem solving. The end of each chapter offers "Applied Problems," so students can gain proficiency and confidence in problem solving. In closing, it was our intent to create a workbook that would serve as a resource to help students master the basics of prealgebra while effectively preparing them for their subsequent algebra courses.

Joshua Yarmish
Rachel Sturm-Beiss

Chapter 1

Integers

SECTION A. Whole Numbers

Let us begin with some basic facts about our number system.

The whole numbers are as follows   0, 1, 2, 3, 4, 5, 6, 7, 8, 9, 10, 11, 12, ...

The three dots indicate that there is no largest number, or that they "go on forever." The symbols 0, 1, 2, 3, 4, 5, 6, 7, 8, and 9 are digits. The value of a digit within a number depends on it's placement.

In the number 534,
the 4 represents 4 ones,
the 3 represents 30 or 3 tens,
the 5 represents 500 or 5 hundreds.

EXAMPLE 1:   In the number 38,675 what do the digits represent?

SOLUTION:
The 5 represents 5 ones.
The 7 represents 7 tens, or 70.
The 6 represents 6 hundreds, or 600.
The 8 represents 8 thousands, or 8000.
The 3 represents 3 ten thousands, or 30,000.

We sometimes write numbers in **expanded notation**. The number 4673 is written in expanded notation as follows.

$$4000 + 600 + 70 + 3$$

A number's **word name** is how the number is read.

EXAMPLE 2:   Write the word name for 254,789.

SOLUTION: Two hundred and fifty-four thousand, seven hundred, eighty-nine.

Addition of Whole Numbers (+)

We add two or more whole numbers by lining the numbers up vertically and adding within each column. If a column sum is 10 or more, we place the right digit of the column sum beneath the column, and "carry" the left digit to the next column.

EXAMPLE 3: Add 548 + 352.

SOLUTION:

$$\begin{array}{r} \overset{1\ 1}{} \\ 548 \\ 352 \\ \hline 900 \end{array}$$

We can reverse the order of addition without affecting the answer. This addition property is called the commutative property of addition.

Commutative Property of Addition: $a + b = b + a$

For example $5 + 4 = 4 + 5$ is an illustration of the commutative property of addition.

If we add three or more numbers, the addition may be grouped in any order. This addition property is called the associative property of addition.

Associative Property of Addition: $(a + b) + c = a + (b + c)$

The following illustration demonstrates an instance where the associative property is used to simplify an addition.

Suppose we are adding the following prices.

$$\begin{aligned} (\$589 + \$395) + \$105 &= \\ \$589 + (\$395 + \$105) &= \qquad \text{This grouping simplifies the addition} \\ \$589 + \$500 &= \\ &= \$1089 \end{aligned}$$

The number 0 is called the **additive identity,** since if we add 0 to any number its value remains the same.

Subtraction of Whole Numbers (–)

We subtract two whole numbers by lining the numbers up vertically with the larger number on top. Starting with the right most column we subtract the lower digit from the upper digit when the upper digit is greater that the lower digit. If the upper digit is less that the lower digit then we "borrow" from the column to the left.

EXAMPLE 4: Subtract 657 – 382.

SOLUTION:

$$\begin{array}{r} {}^{5}\!\!\!{}^{1} \\ \cancel{6}\,5\,7 \\ -\,3\,8\,2 \\ \hline 2\,7\,5 \end{array}$$

Multiplication of Whole Numbers (×, •, juxtaposition)

Numbers that are multiplied are called **factors**, and the result of a multiplication is a **product**. For example $2 \times 3 = 6$; 2 and 3 are factors (of 6) and 6 is the product of 2 and 3.

Unlike addition and subtraction, there are many ways to indicate multiplication.
The product 2 times 3 can be expressed as:

2×3	multiplication operator notation
$2 \cdot 3$	dot notation
$(2)(3)$	juxtaposition
$2(3)$	juxtaposition
$(2)3$	juxtaposition

In this presentation we assume the reader knows the multiplication tables for all single digit whole numbers.

Multiplication of a whole number times 10, adds a zero onto the end of the number. Multiplication by 100 adds two zeros onto the end, and multiplication by 1000 adds three zeros onto the end of the whole number being multiplied.

EXAMPLE 5: Multiplication by 10, 100 and 1000.
    (a) 10×456 **(b)** 100×456 **(c)** 1000×456

SOLUTION:
    (a) $10 \times 456 = 4560$
    (b) $100 \times 456 = 45,600$
    (c) $1000 \times 456 = 456,000$

3

Multiplication of a one digit number by a number whose digits are all zero except for the leading (left-most) digit is illustrated here.

$$400 \times 2 =$$
$$(4 \times 2)(100) = \quad \text{Multiply the leading digits by the whole number formed from a}$$
$$8 \times 100 = \quad \text{1 followed by the number of zeros that are in the first number.}$$
$$= 800$$

EXAMPLE 6: Multiply.

    (a) 7000×4 **(b)** $30,000 \times 2$

SOLUTION:

    (a) $7000 \times 4 = 28,000$

    (b) $30,000 \times 2 = 60,000$

Now, suppose we want to multiply any whole number by a one digit number. For example,

```
 1 1
 367
×  2
 734
```

    Multiply $2 \times 7 = 14$, put the 4 in the right-most place and "carry" the 1.
    Multiply $2 \times 6 = 12$, and add the carried 1 to get 13. Put the 3 next to
        the 4 and "carry" the 1.
    Multiply $2 \times 3 = 6$, and add the carried 1 to get 7. Put the 7 next to the 3.

More generally, suppose we want to multiply the whole numbers 367×42.

```
  367
 ×42
  734
 1468
15414
```

    734   Multiply 2×367, as in the example above.
    1468   Multiply 4×367, place the result under the first product, shifted one place left.
    15414 Sum the rows above.

EXAMPLE 7: Multiply 563×892.

SOLUTION:

```
   563
  ×892
  1126
  5067
  4504
502196
```

We can reverse the order of a multiplication problem without affecting the solution. This property of multiplication is called the commutative property.

The Commutative Property of Multiplication: $a \times b = b \times a$

For example, we have $3 \times 5 = 5 \times 3$.

If we multiply three or more numbers the multiplication can be grouped in any order. This property is called the associative property of multiplication.

The Associative Property of Multiplication: $(a \times b) \times c = a \times (b \times c)$

Multiplication by the number 1 does not change the value of the number being multiplied. Thus, the number 1 is called the **multiplicative identity**.

The following property, called the distributive property, tells us how to multiply a number, times the sum of two numbers.

The Distributive Property: $a(b + c) = ab + ac$

For example $5(2 + 7) = 5(2) + 5(7) = 10 + 35 = 45$. Notice that if we do the operation in parenthesis first, we still get 45: $5(2 + 7) = 5(9) = 45$.

Division of Whole Numbers $\left(\div, /, \overline{)} \right)$

The number 15 is equal to 3×5. Thus, we can divide 15 items into 3 groups with 5 items in each group. If we write $15 = 5 \times 3$ then we see that 15 items can also be divided into 5 groups with 3 items in each group. We say that $15 \div 5 = 3$ (15 divided by 5 equals 3) and $15 \div 3 = 5$. In the problem $15 \div 5 = 3$, we call 15 the **dividend**, 5 the **divisor**, and 3 the **quotient**.

There are many symbols used to indicate division. The following expressions all mean $15 \div 5$.

$$15 \div 5 \qquad 15/5 \qquad \frac{15}{5} \qquad 5\overline{)15}$$

Simple division problems can be done with "mental arithmetic". Here are some examples.

$14 \div 2 = 7$ since $2 \times 7 = 14$

$35 \div 7 = 5$ since $7 \times 5 = 35$

$27 \div 3 = 8$, Remainder 3   since $3 \times 8 = 24$, and $24 + 3 = 27$

5

We perform "long division" for division problems that involve larger numbers. For example,

$$\begin{array}{r} 1 \\ 5\overline{)723} \\ 5\downarrow \\ \hline 22 \end{array}$$

First divide 7 by 5; $7 \div 5 = 1$ (remainder 2). Put the 1 in the quotient.
Multiply 1×7 and place the product under the 7.
Subtract, and "bring down the 2"

$$\begin{array}{r} 14 \\ 5\overline{)723} \\ 5 \\ \hline 22 \\ 20\downarrow \\ \hline 23 \end{array}$$

Divide 22 by 5; $22 \div 5 = 4$ (remainder 2). Put the 4 in the quotient.
Multiply 4×5 and place the product of 20 under the 22.
Subtract, and "bring down the 3"

$$\begin{array}{r} 144 \\ 5\overline{)723} \\ 5 \\ \hline 22 \\ 20 \\ \hline 23 \\ 20 \\ \hline 3 \end{array}$$

Divide 23 by 5; $23 \div 5 = 4$ (remainder 3). Put the 4 in the quotient.
Multiply 4×5 and place the product of 20 under the 23.
Subtract; there are no digits left to "bring down." Three is the remainder.

The quotient is 144 remainder 3.

EXAMPLE 8:　$356 \div 25$　　Divide and check.

SOLUTION:

$$\begin{array}{r} 14 \\ 25\overline{)356} \\ 25 \\ \hline 106 \\ 100 \\ \hline 6 \end{array}$$

The quotient is 14 remainder 6.

The answer is correct if (divisor × quotient) + remainder = dividend .

<u>Check</u>: Multiply $25 \times 14 = 350$; add the remainder $350 + 6 = 356$.

Exercises: Whole Numbers

For 1 – 4, state what the digits represent within each number.

1. 453   **2.** 1,987   **3.** 12,345   **4.** 167,543

For 5 – 8, write in expanded form.

5. 345   **6.** 1,378   **7.** 13,456   **8.** 789,345

For 9 – 12, write the word name for the given number.

9. 576   **10.** 12,567   **11.** 567,345   **12.** 1,234,789

For 13 – 15, add.

13. 478 + 678   **14.** 9874 + 478   **15.** 5080 + 2456

For 16 – 18, subtract.

16. 2455 – 1349   **17.** 789 – 298   **18.** 1230 – 986

For 19 – 30, multiply.

19. 10(987)   **20.** 100×5670 **21.** 1000(56)

22. 400(3)   **23.** 7000(8)   **24.** 30,000(9)

25. 3(456)   **26.** 45(287)   **27.** (452)23

28. 783×45 **29.** (563)(38)   **30.** 678×347

31. Use the associative property to simplify the following addition: 789 + 105 +195.

For 32 – 33, expand using the distributive property.

32. 6(7 + 8)   **33.** 5(12 + 20)

7

For 34 – 39, divide and check. State the quotient and remainder.

34. $789 \div 45$ **35.** $7865 \div 50$ **36.** $345 \div 23$

37. $582 \div 32$ **38.** $7234 \div 40$ **39.** $345 \div 10$

40. Find the product of 23 and 47.

41. How many weeks are in 210 days?

42. How many days are in 54 weeks?

43. If fifty-six dollars is divided evenly among 8 people, then each person receives _____ dollars.
(a) 6   (b) 9   (c) 7   (d) 4   (e) 8

44. If 63 chairs are arranged in a formation of 9 rows, then each row has _____ chairs.
(a) 6   (b) 7   (c) 8   (d) 9   (e) 5

45. If 77 pieces of candy is split evenly among 11 students, then each student receives _____ pieces.
(a) 10   (b) 11   (c) 20   (d) 7   (e) 9

46. If 68 pieces of candy is split evenly among 12 people, then how many pieces will be left over?
(a) 8   (b) 5   (c) 12   (d) 10   (e) 6

47. How many square feet are in 13 square yards?
(a) 39   (b) 169   (c) 52   (d) 156   (e) 117

48. A merchant paid $54 dollars for 9 yards of material. How much did he pay for one foot of material?
(a) $2   (b) $6   (c) $18   (d) $3   (e) $9

8

49. If 15 square yards of carpeting were ordered to cover 130 square feet of floor space, how many square feet of carpeting will be left after the floor is covered.
　　(a) 10　　　(b) 20　　　(c) 3　　　(d) 5　　　(e) 0

50. A tablecloth is 104 inches covers a 7 foot long table. How many inches of cloth hang over the front, assuming the table is centered in the tablecloth?
　　(a) 20　　　(b) 5　　　(c) 10　　　(d) 34　　　(e) 15

51. A man bought a television for $490. He gave a deposit of $100 and paid the rest in 10 equal installments. How much was each installment?
　　(a) 49　　　(b) 19　　　(c) 59　　　(d) 39　　　(e) 10

52. A DVD plays for 95 minutes. If it starts at 1:05 PM, what time will it end?
　　(a) 2:35 PM　(b) 2:40 PM　(c) 2:15 PM　(d) 3:00 PM　(e) 2:45 PM

For 53 – 54, the four members of the Thomas family are planning to drive from Baltimore, Maryland to Los Angeles, California, a trip of 3360 miles. They estimate that the trip will take 2 weeks.

53. If they cover the same number of miles each day, how far will they drive in a single day?
　　(a) 200　　　(b) 336　　　(c) 350　　　(d) 340　　　(e) 240

54. Suppose the family drives at a steady speed of 48 miles per hour. How many hours do they drive each day?
　　(a) 5　　　(b) 8　　　(c) 6　　　(d) 12　　　(e) 7

55. A builder must buy enough tiles to cover 120 sq ft of wall. The tiles are 8 in. by 8 in.. How many tiles does he need?
　　(a) 15　　　(b) 270　　　(c) 1875　　　(d) 23　　　(e) 24

56. What number added to 34 gives 123?
　　(a) 77　　　(b) 157　　　(c) 89　　　(d) 79　　　(e) 67

57. How much more than 65 is 342?
　　(a) 277　　　(b) 255　　　(c) 407　　　(d) 177　　　(e) 292

9

SECTION B. Introducing Integers

The integers are the numbers   ...., -3, -2, -1, 0, 1, 2, 3, ...

The numbers to the right of 0 are the positive numbers and are greater than 0. The numbers to the left of 0 are the negative numbers and are less than 0. Notice that 0 is neither positive nor negative. The numbers increase from left to right. We use the symbol "–"to indicate that a number is negative and the symbol "+" to indicate that a number is positive. Thus the number –5 represents "negative five", and the number +5 or just 5 represents "positive five."

We often come across situations that require us to consider numbers that are less than 0. For example, on a thermometer a temperature of 10 degrees below zero is represented as –10°.

Here are some more real-life examples of negative quantities:

1. If your checking account has $10 in it and you write a check for $15, you can represent the balance as –$5.
2. The peak of Mt. Everest, Nepal is 29,028 ft. above sea level. The deepest point of the Gulf of Mexico is 14,370 ft. below sea level. You can represent the depth as –14,370 ft.
3. If + $25 stands for a gain of $25 then –$25 stands for a loss of $25.

EXAMPLE 1:   Representing negative quantities
    a. If +4 inches means 4 inches above average, then what does –3 inches mean?
    b. If North 150 miles is represented by +150, what does –150 represent?
    c. What is a *gain* of –$25?

SOLUTION:
    a. Three inches below average.
    b. South 150 miles.
    c. A *loss* of $25.

We can also picture integers using a number line.

Numbers increase in value as we move from left to right on the number line. Therefore –3 is less than 1 (–3 < 1), and –5 is less than –3 (–5 < –3).

EXAMPLE 2: Replace the ? with the inequality symbol < or >.
  a. −7 ? 2
  b. 4 ? −7
  c. −8 ? −2

SOLUTION:
  a. −7 < 2
  b. 4 > −7
  c. −8 < −2

EXAMPLE 3: Write the numbers in increasing order.
    −7, 10, 0, −6, 3, 2, −4

SOLUTION:
    −7, −6, −4, 0, 2, 3, 10

Numbers that are the same distance from zero but are on opposite sides of zero are called **opposites**. For example, 3 and −3 are opposites.

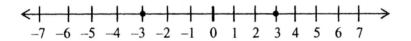

EXAMPLE 4: State the opposite.
  a. −5
  b. 7

SOLUTION:
  a. 5
  b. −7

11

We can also find the opposite of a number by placing a negative symbol in front of the number. The opposite of 2 is –2 and the opposite of –3 is –(–3), which simplifies to +3 or 3.

EXAMPLE 5: Simplify.

      a. –(–5)
      b. –(–(–13))

SOLUTION:

      a. 5
      b. –(–(–13))= –(13) = –13

EXAMPLE 6: Let $S = \{1, 2, 3, 4, 5, 6, \ldots, 48\}$
      Let $T = \{$all elements of S, and the opposites of all the elements in $S\}$.
      Arrange the elements of T in 6 columns:

$$
\begin{array}{cccccc}
1 & 2 & 3 & 4 & 5 & 6 \\
7 & 8 & 9 & 10 & 11 & 12 \\
\vdots & \vdots & \vdots & \vdots & \vdots & \vdots
\end{array}
$$

      How many rows are there?

SOLUTION:

      Since T has $48 + 48 = 96$ elements, and $(6)(16) = 96$. Therefore there are 16 rows.

SECTION C. Absolute Values

The absolute value of an integer is the value of the number without regard to its sign. The absolute value of +7 is 7 and the absolute value of –5 is 5.

The absolute value of –5 is represented by $|-5|$, where $|-5| = 5$.

EXAMPLE 1: Simplify the expressions involving absolute values.
  a. $|-7|$
  b. $|3|$
  c. $-|4|$
  d. $-|-9|$

SOLUTION:
  a. $|-7| = 7$
  b. $|3| = 3$
  c. $-|4| = -4$
  d. $-|-9| = -9$

The absolute value of a number is also the distance between that number and 0 on the number line.

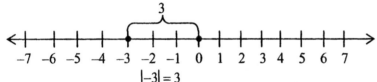

$$|-3| = 3$$

Exercises: Introducing Integers and Absolute Values

1. If 4° above zero is represented by +4°, then how do we represent 3° below zero?

2. If +1500 ft represents an elevation of 1500 ft above sea level, then what does −1105 ft represent?

3. Explain what a "gain" of −$5 is.

4. What does −35 miles North mean?

5. Which is colder: −5° or −9°?

6. Which is colder: 4° or −8°?

For 7 – 9, replace the ? with the inequality symbol < or >.

7. 7 ? −5

8. −3 ? −2

9. 4 ? −10

For 10 – 11, arrange the integers in increasing order.

10. −4, 9, 8, 0, −3, −2, 5, −7

11. −5, 6, 8, −3, 0, −2, 3

For 12 – 13, given a number, find its opposite.

12. 6

13. −5

For 14 – 15, simplify.

    14. –(–9)

    15. –(–(–6))

    16. What is the distance on the number line, between 7 and its opposite?

For 17 – 19, state the absolute value.

    17. | 3 |

    18. |–3|

    19. |–12|

For 20 – 22, replace the ? with the appropriate inequality symbol < or >.

    20. |–3| ? | 2 |

    21. |–1| ? –| 5 |

    22. –|–13| ? |–4|

    23. The set S = {10, 20, 30, 40, …, 90}.
    Let T = {elements of S and the opposites of all the elements of S}. If set T were arranged
    in rows with 3 elements in each row, how many rows would there be?

15

SECTION D. Operations with Integers

Adding Integers

John is playing black jack. In the first round, he wins $10. In the next round he wins $5. We associate a gain with a positive symbol (+), and express his total gain as
$$(+\$10) + (+\$5) = +\$15$$

Suppose John loses $10 in the first round and loses $5 in the second round. His total loss is then $15. We associate a loss with a negative symbol (–), and express his total negative gain as
$$(-\$10) + (-\$5) = -\$15$$

If John wins $10 in the first round and loses $15 in the second, then his total would be a loss of $5. We represent his negative gain as
$$(+\$10) + (-\$15) = -\$5$$

And finally, if John wins $20 in the first round and loses $15 in the second round, his total gain is $5. We represent his positive gain as
$$(+\$20) + (-\$15) = +\$5$$

EXAMPLE 1:   Think of the positive numbers as gains and the negative numbers as losses. Combine the following:
   a. $(-8) + (-13)$
   b. $(+25) + (-12)$
   c. $(+14) + (-27)$

SOLUTION:
   a. $(-8) + (-13) = -21$
   b. $(+25) + (-12) = +13$
   c. $(+14) + (-27) = -13$

We can use the number line to help find the sum of two integers. A move to the right on the number line is a move in the positive direction, and a move to the left is a move in the negative direction. We find the sum of $(-3) + 5$ by starting at zero and moving 3 units in the negative direction (left) followed by 5 units in the positive direction (right). We end at 2, which is the sum.

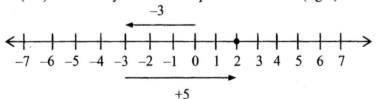

16

EXAMPLE 2: Use the number line to find the sums.
   a. −2 + (−3)
   b. 4 + (−3)
   c. 3 + (−4)

SOLUTION:
   a. −2 + (−3) = −5

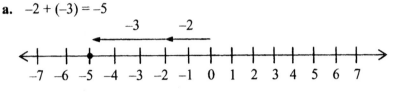

   b. 4 + (−3) = 1

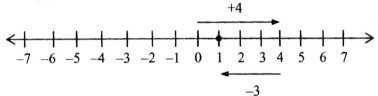

   c. 3 + (−4) = −1

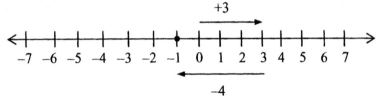

Rules for adding integers:
1. To add two numbers with like signs, use the same sign and find the sum of their absolute values.
2. To add two numbers with different signs, use the sign of the number with larger absolute value and find the difference of their absolute values.

EXAMPLE 3: Combine.
   a. (−56) + (−72)
   b. (−78) + (26)
   c. (32) + (−15)

SOLUTION:
   a. (−56) + (−72) = −128
   b. (−78) + (26) = −52
   c. (32) + (−15) = 17

17

EXAMPLE 4: Using positive and negative numbers, rewrite the following and then combine.
   a. Losing $6 and gaining $5.
   b. Making a profit of $80 and a loss of $60.

SOLUTION:
   a. $(-6) + (5) = -1$
   b. $(80) + (-60) = 20$

EXAMPLE 5: Adding more than two integers.
$$-5 + 3 + (-7) + 4 =$$

SOLUTION: The commutative property of addition tells us that we can change the order of addition. We first add the negative numbers, and then we add the positive numbers:
$$-5 + 3 + (-7) + 4 =$$
$$-5 + (-7) + 3 + 4 =$$
$$-12 \quad + 7 \ = -5$$

Subtracting Integers

Kin has $50 in the bank. He writes a check for $100. If the bank cashes the check, then Kin will have $50 - $100 = -$50.

Suppose Kin's bank account has an overdraft of $50 (a balance of -$50) and he writes a check for $25, then Kin will have -$50 - $25 = -$75.

Arnie's online store has 5 pepper mills in stock. A customer places an order for 6 pepper mills. He is short one mill, or has -1 mills left. Thus $5 - 6 = -1$.

Notice that "five subtract six" is $5 - 6 = -1$.
Notice that "five plus negative six" is $5 + (-6) = -1$.
Both have the same result, and we can write" five plus negative six" as $5 + (-6)$ or as $5 - 6$. You will soon realize that *adding a negative number is equivalent to subtracting a positive number.*

Rule for Subtracting Two Numbers:
$$a - b = a + (-b)$$
To subtract, add the opposite of the second number to the first.

EXAMPLE 6: Subtract.
  a. $-5 - 2$
  b. $-4 - (-6)$

SOLUTION:
  a. $-5 - 2 = -5 + (-2) = -7$
  b. $-4 - (-6) = -4 + 6 = 2$

EXAMPLE 7: Interpret as a subtraction problem and solve.
  a. Find the difference between -17 and 35.
  b. Find the difference between a gain of $23 and a loss of $18.
  c. Suppose a loss of $5 was mistakenly entered in a check book. If the (incorrect) balance is $83 find the correct balance by subtracting the $5 loss.
  d. The Lake Placid weather station registered a temperature of $-12°F$ at 5:00 AM. At noon the temperature was $15°F$. How many degrees did the temperature rise?
  e. Subtract 23 from -16.
  f. The Dead Sea has an altitude of 800 meters below sea level, or -800 meters. Mount Everest has an altitude of 8872 meters above sea level. What is the difference between the altitude of Mount Everest and that of the Dead Sea?

SOLUTION:
  a. $-17 - 35 = -17 + (-35) = -52$
  b. $23 - (-18) = 23 + 18 = \$41$
  c. $83 - (-5) = 83 + 5 = \$88$
  d. $15 - (-12) = 15 + 12 = 27°$
  e. $-16 - 23 = -16 + (-23) = -39$
  f. $8872 - (-800) = 8872 + 800 = 9672$ meters

Multiplying Integers

There are four types of products. These are: a positive number times a positive number, a positive number times a negative number, a negative number times a positive number and a negative number times a negative number.

A positive times a positive. Suppose you save $10 a week, then in 5 weeks you will be $50 richer. Thus, $(5) \times (\$10) = \$50$.

A positive times a negative. Suppose you lose $10 a week, then in 5 weeks you will be $50 poorer. Thus, $(5) \times (-\$10) = -\$50$.

19

A negative times a positive. Suppose you save $10 a week, then 5 weeks ago you were $50 poorer. Thus, $(-5) \times (\$10) = -\$50$.

A negative times a negative. Suppose you lose $10 a week. Then 5 weeks ago you were $50 richer. Thus, $(-5) \times (-\$10) = +\$50$.

Rules for Multiplication of Integers:
1. The product of two numbers with like signs is positive.
2. The product of two numbers with unlike signs is negative.

EXAMPLE 8: Multiply.
    a. $(5)(-7)$
    b. $-6(5)$
    c. $(-12) \times (-6)$

SOLUTION:
    a. $(5)(-7) = -35$
    b. $-6(5) = -30$
    c. $(-12) \times (-6) = 72$

EXAMPLE 9: Multiply four numbers
    $(3)(-6)(5)(-2) =$

SOLUTION:
    $(3)(-6)(5)(-2) =$
    $(-18)(5)(-2) =$
    $(-90)(-2) = +180$

EXAMPLE 10: Interpret as a multiplication problem and solve.
In math class, Mr. Baker lowered a student's grade 3 points for every missed assignment. The student missed 5 assignments. Express the number of points in the student's grade change as an integer.

SOLUTION: $(5)(-3) = -15$

Dividing Integers

We know that if a number is a product of two numbers, and is divided by either of them, then the quotient is the other number. We can learn the rules for division of integers from the following examples.

Since $(+5)(+3) = (+15)$, then $(+15) \div (+3) = +5$.
Since $(-5)(+3) = (-15)$, then $(-15) \div (+3) = -5$.
Since $(+5)(-3) = (-15)$, then $(-15) \div (-3) = +5$.
Since $(-5)(-3) = (+15)$, then $(+15) \div (-3) = -5$.

These are examples of the following rules.

Rules for dividing integers:
1. The quotient of two numbers with like signs is positive.
2. The quotient of two numbers with unlike signs is negative.

EXAMPLE 11: Divide
    a. $-28 \div 7$ **b.** $35 \div (-7)$ **c.** $-45 \div (-9)$

SOLUTION:
    a. $-28 \div 7 = -4$
    b. $35 \div (-7) = -5$
    c. $-45 \div (-9) = +5$

We cannot divide by zero. Suppose we divided the number 5 by 0. The result multiplied by zero would have to be 5. This is impossible since any number multiplied by 0 is 0.

EXAMPLE 12: Interpret as a division problem and solve.
Ever since the opening of a new shopping center, Mr. Brown's clothing store has lost revenue. If the shopping center opened 6 months ago and Mr. Brown's store earned a negative profit of $-\$480$ over the past 6 months, then on the average what was his monthly (negative) profit?

SOLUTION: $-\$480 \div 6 = -\$80$

Exercises: Operations with Integers

For 1 – 3, think of the positive numbers as gains and negative numbers as losses. Combine the
following.

   1. $-5+(10)$ **2.** $-9+(-5)$ **3.** $13+(-4)$

For 4 – 6, use the number line to find the sums.

   4. $4+(-3)$ **5.** $-2+(-3)$ **6.** $3+(-5)$

For 7 – 18, perform the indicated operation.

   7. $-89+36$ **8.** $-34+67$ **9.** $-73+(-98)$

  10. $88-97$ **11.** $54-35$ **12.** $94-(-76)$

  13. $-13-(-39)$ **14.** $56-(-77)$ **15.** $25-13$

  16. $-87+(-43)$ **17.** $-17-(-65)$ **18.** $53-98$

For 19 – 26, perform the indicated operation.

  19. $(9)(-13)$ **20.** $-7(5)$ **21.** $(-32)(-3)$

  22. $(2)(-6)(4)(-6)$ **23.** $-85\div5$ **24.** $-21\div(-7)$

  25. $45\div(-9)$ **26.** $-121\div(-11)$

  27. Subtract 34 from -17.

  28. Divide -104 by 13.

  29. Joe loses $4 each week for a period of 6 weeks. Express his loss as an integer.

30. The famous cliff divers of La Quebrada, Mexico, jump into the ocean from a cliff 113 feet above sea level and plunge to a depth of 60 feet below sea level, (- 60 feet). In total, how many feet do they descend?
  (a) 109   **(b)** 53   **(c)** 50   **(d)** 173   **(e)** 115

31. Suppose a checking account balance appears to be $100. If a gain of $5 was mistakenly entered then the correct balance is $100 - 5 = \$95$. Suppose instead that a *loss* of $5 was mistakenly entered. Then the loss is represented by –$5, and the correct balance is $100 - (-5) =$ _____
  (a) 115   **(b)** 95   **(c)** 110   **(d)** 90   **(e)** 105

32. In a football game the Giants had the ball on their opponents' 14 yard line. They made a gain of 7 yards (+7) on one play and then received a 5-yard penalty (–5). Where was the ball placed after the penalty?
  (a) 17   **(b)** 26   **(c)** 12   **(d)** 16   **(e)** 30

33. What is the difference in degrees between 6° below zero and 18° below zero?
  (a) 24   **(b)** –22   **(c)** 12   **(d)** –12   **(e)** 22

34. The temperature at dawn in Lake Placid was -20°, and at noon it was 18°. By how many degrees did the temperature rise?
  (a) 28   **(b)** 18   **(c)** 38   **(d)** 2   **(e)** –2

35. Joe won $30 at black jack and then lost $43 at the slot machines. What are his net earnings? Express your answer as a negative number.
  (a) –$23   **(b)** –$73   **(c)** –$43   **(d)** –$30   **(e)** –$13

36. Kim won $45 dollars at black jack and then lost $40 playing the slot machines. Kelly won $33 dollars at the roulette wheel and then lost $40 playing the slot machines. What is the difference between Kim's earnings and Kelly's (negative) earnings?
  (a) $12   **(b)** $2   **(c)** –$2   **(d)** $10   **(e)** –$10

37. Kim took the SAT exam twice. The first time she received a score of 1050, and the second time she received a score of 1173. Kim's friend got a first score of 1132 and a second score of 1195. Kim's score improved by how many more points than her friend's?
  (a) 145   **(b)** 123   **(c)** 33   **(d)** 60   **(e)** 156

38. Ted spends $6 every week day (Monday - Friday) on Lottery games. The previous year he won a total of $454. What was his net gain last year?
  (a) –$1234   **(b)** –$1371   **(c)** $1234   **(d)** –$1106   **(e)** $946

23

SECTION E. Order of Arithmetic Operations and Exponents

Exponents

The expression $2 \times 2 \times 2$ can be written as 2^3. This is read "2 to the 3rd power" or "2 raised to an exponent of 3" or "2 cubed." The number 2 is called the **base** and the number 3 is called the **exponent**.

EXAMPLE 1: Write in exponent form.
  a. (4)(4)(4)
  b. (−7)(−7)(−7)(−7)

SOLUTION:
  a. 4^3
  b. $(-7)^4$

EXAMPLE 2: Write in expanded form.
  a. 3^5
  b. $5x^3$

SOLUTION:
  a. $3^5 = 3 \cdot 3 \cdot 3 \cdot 3 \cdot 3$
  b. $5x^3 = 5 \cdot x \cdot x \cdot x$ Notice that the exponent of 3 applies only to the x and not to the 5.

Order of Operations

Suppose a group of students were asked to find the value of $3 + 4 \times 5$. Some students might multiply 4×5 first and get 20, and then add $3 + 20$ and get 23. Others might add $3 + 4$ first and get 7 and then multiply 7×5 and get (the wrong answer) 35. In order that the expression $3 + 4 \times 5$ has only one value we must agree on the order of performing the indicated operations. The following rules state the agreed upon order of performing operations.

Order of Operations
1. First do operations inside parenthesis.
2. Next simplify any expressions with exponents.
3. Next do multiplications or divisions from left to right.
4. Next do additions or subtractions from left to right.

EXAMPLE 3:   Perform the indicated operations.

    a. $5+3(4)$ **b.** $6\times(-3)-5\times4$

    c. $4-5(3-12)$ **d.** $(5)(2)^3$

    e. -2^4 **f.** $3-2[6-(7-3)]$

    g. $(-3)(2)+\dfrac{4(3-(-2))}{2(2-(-3))}$

SOLUTION:

    a. $5+3(4)=$
            $5+12=17$

    b. $6\times(-3)-5\times4=$
            $-18\ -\ 20=-38$

    c. $4-5(3-12)=$
            $4-5(-9)\ =$
              $4-(-45)=$
               $4+45=49$

    d. $(5)(2)^3=$
            $(5)(8)=40$

    e. $-2^4=$
       $-2\cdot2\cdot2\cdot2=-16$

    f. $3-2[6-(7-3)]=$
        $3\ -\ 2[6-4]\ =$
          $3\ -2[2]\ =$
            $3-4=-1$

    g. $(-3)(2)+\dfrac{4(3-(-2))}{2(2-(-3))}=$
        $-6\ +\ \dfrac{4(3+2)}{2(2+3)}=$
        $-6\ +\ \dfrac{20}{10}\ =$
          $-6+2=-4$

EXAMPLE 4:   Subtract $8\cdot3-4$ from $8\cdot(3-4)$

SOLUTION: $8\cdot(3-4)-[8\cdot3-4]=$
           $8\cdot(-1)-[24-4]=$
           $8\cdot(-1)-[20]=$
            $-8-20=-28$

Exercises: Order of Arithmetic Operations and Exponents

For 1 – 7, perform the indicated operation.

1. $4 + 5 \times 3$

2. $6 - 4(3 + 7)$

3. $9 - 2(4 - 8)$

4. $4(5) - 3(2)$

5. $8 - (4 + 3) \times 2 - 4$

6. $4 - 5 + 7(4) - 5$

7. $4 - 9 \div 3(3) + 5$

For 8 – 10, write in exponent form.

8. $(5)(5)(5)$

9. $(-2)(-2)(-2)(-2)$

10. $(10)(3)(3)$

For 11 – 14, write in expanded form.

11. $(2)^3$

12. $5(3)^4$

13. $(3)^2(7)$

14. $5x^4$

For 15 – 20, perform the indicated operations.

15. $(7)(2)^2$

16. $1 - 3^2$

17. -3^2

18. $(-3)^2$

19. $-(-2)^3$

20. $-3^2 2^3$

21. Subtract $6(7) - 5$ from $4 - (5)(3)$.

22. Subtract $-8 - (3)(5)$ from $(8-3)(5)$.

23. $(-3)(5) + \dfrac{3(5 - (-2))}{-3(4 + 3)}$

26

SECTION F. Multiples and Factors of Integers

The number 3 is a **factor** of 6, since 3 divides 6 "evenly" (without a remainder). We can write 6 as the product (3)(2). The word "factor" is *also* used as a verb, as in: we factor 6 as (2)(3). We also say that (2)(3) is a **factorization** of 6.

Note: Since a factor of a number divides into the number evenly, a factor is also called a **divisor**.

The number 12 can be factored as (1)(6) or (4)(3) or (2)(6) or (2)(2)(3). The numbers 4, 3, 2, and 6, are all factors of 12. Notice that the last factorization has more that two factors.

A number is called a **prime number** if it is greater than 1, and if its only factors are 1 and itself. For example, the number 4 is not a prime number since $4 = (2)(2)$, and the number 5 is a prime number since $5 = (1)(5)$ is the only factorization of 5.

EXAMPLE 1:
    a. List all the two-factor products of 18.
    b. Write 18 as a product of prime factors.

SOLUTION:
    a. (1)(18), (2)(9), (6)(3)
    b. $2 \cdot 9$
          /\
      $2 \cdot 3 \cdot 3$
      $2 \cdot 3^2$ the prime factorization

Every number has only one prime factorization. This is called the *fundamental theorem of arithmetic.*

Fundamental Theorem of Arithmetic: Every integer greater than 1 is either prime or can be expressed as a product of prime factors. Except for the order of the factors, this can be done in one and only one way.

EXAMPLE 2: Write the product of prime factors.
    a. 24
    b. 30

SOLUTION: **a.**   24   **b.**   30
                  /\   /\
                $6 \cdot 4$ $6 \cdot 5$
              /\ /\   /\ \
            $3 \cdot 2 \cdot 2 \cdot 2$ $3 \cdot 2 \cdot 5$
              $3 \cdot 2^3$

27

The Greatest Common Divisor

The numbers 12 and 18 are both divisible by 1, 2, 3, and 6. The largest number that divides both is called the greatest common divisor. Thus, 6 is the greatest common divisor of 12 and 18.

The **greatest common divisor (GCD)** of two integers is the largest integer that divides the two integers evenly (without a remainder).

Note: The greatest common divisor is also called the **greatest common factor**.

EXAMPLE 3:   Find the GCD of 30 and 45.

SOLUTION:
The divisors of 30 are: 1, 2, 3, 5, 6, 10, 15, and 30.
The divisors of 45 are: 1, 3, 5, 9, 15, and 45.
The GCD is 15.

We can also find the GCD of two numbers from their prime factorizations. The GCD contains the prime factors that are common to both numbers' prime factorizations. Each prime factor appears the least number of times it occurs in each number.

EXAMPLE 4:   Find the GCD of 18 and 60 from the prime factorizations of both numbers.

SOLUTION:
The prime factorization of $18 = 2 \cdot 3 \cdot 3$.
The prime factorization of $60 = 2 \cdot 2 \cdot 3 \cdot 5$.
The primes 2, and 3 appear in both numbers. The prime 2, appears once in 18 and twice in 60, so we include it once in the GCD. The prime 3 appears twice in 18 and once in 60, so we include it once in the GCD.
The GCD is $2 \cdot 3 = 6$.

The Least Common Multiple

The number 6 is a **multiple** of 3 since $6 = (3)(2)$. Stated differently, 3 divides 6 evenly. Some more multiples of 3 are 9, 12, 15, 18, 21, 24, etc.

EXAMPLE 5:   List ten multiples of 5.

SOLUTION:
5, 10, 15, 20 , 25, 30, 35, 40, 45, 50   Notice that 5 is a multiple of itself sine $5 = (5)(1)$.

EXAMPLE 6: List three multiples of both 6 and 9.

SOLUTION: 18, 36, 54

The **least common multiple (LCM)** of two integers, is the smallest positive integer that is a multiple of both integers.

EXAMPLE 7: Find the LCM of 12 and 18.

SOLUTION:
The multiples of 12 are: 12, 24, <u>36</u>, 48, …
The multiples or 18 are: 18, <u>36</u>, 54, 72, ...
The LCM is 36.

We can also find the LCM of two numbers from their prime factorizations. The LCM is the product of the different prime factors of each number, each prime appearing the greatest number of times it occurs in each number.

EXAMPLE 8: Find the LCM of 36 and 120 from their prime factorizations.

SOLUTION:
The prime factorization of 36 is $2 \cdot 2 \cdot 3 \cdot 3$.
The prime factorization of 120 is $2 \cdot 2 \cdot 2 \cdot 3 \cdot 5$.
The LCM contains all the prime factors: 2, 3, and 5. The prime 2 appears twice in 36 and three times in 120, so it appears three times in the LCM. The prime 3 appears twice in 36 and once in 120, so it appears twice in the LCM. The prime 5 appears only once in 120, so it appears once in the LCM. Therefore the LCM is $2 \cdot 2 \cdot 2 \cdot 3 \cdot 3 \cdot 5 = 360$.

EXAMPLE 9: Karen gets paid every 3 weeks, and she pays her bills every 4 weeks. On which weeks does she get paid and pay her bills?

SOLUTION:
She pays her bills and gets paid on weeks that are multiples of 3 and 4.
Find the LCM of 3 and 4.
Multiples of 3: 3, 6, 9, <u>12</u>, 15, ...
Multiples of 4: 4, 8, <u>12</u>, 16, 20,...
The LCM is 12.
The weeks that are multiples of 12 are multiples of 3 *and* multiples of 4. She gets paid and pays her bills on weeks that are multiples of 12.

EXAMPLE 10: If 15 divides integer x evenly, and 12 divides x evenly, then show that the LCM of 15 and 12 must divide x evenly.

SOLUTION:

If 15 divides x, then x has factors 3 and 5.

If 12 divides x, then x has factors 2^2 and 3.

So, x must have factors 3, 5, and 2^2.

The LCM of 15 and 12 is $3 \cdot 5 \cdot 2^2 = 60$. So, 60 divides x.

If an integer x is divisible by integers n and m, then x is divisible by the LCM of n and m.

EXAMPLE 11: If 15 divides integer x evenly, and 12 divides x evenly, then
  a. must 9 divide x?
  b. must 30 divide x?
  c. must 40 divide x?

SOLUTION:
  a. Nine does not divide the LCM of 60, so it does not necessarily divide x.
  b. Thirty divides 60, so 30 does divide x.
  c. Forty does not divide 60, so it does not necessarily divide x.

Exercises: Multiples and Factors of Integers

For 1 – 4, state all the factors (divisors) of the given number.

   1. 5

   2. 9

   3. 12

   4. 42

   5. What is a prime number?

For 6 – 9, give the prime factorization of the given number.

   6. 3

   7. 15

   8. 21

   9. 42

For 10 – 15, find the greatest common divisor of the pair of numbers.

   10. 9 and 3   **11.** 15 and 10

   12. 12 and 18   **13.** 75 and 30

   14. 36 and 60   **15.** 125 and 75

31

For 16 – 21, find the least common multiple of the pair of numbers.

    16. 5 and 7

    17. 3 and 7

    18. 7 and 14

    19. 21 and 14

    20. 30 and 42

    21. 18 and 30

    22. If a number is divisible by 21 and 6, then it may not be divisible by _____.
       (a) 14   **(b)** 7   **(c)** 2   **(d)** 9   **(e)** 14 and 9

    23. If a number is divisible by 12 and 15, then it may not be divisible by _____.
       (a) 20   **(b)** 6   **(c)** 4   **(d)** 6 and 20   **(e)** 9

    24. A pack of gum has 6 pieces. What is the fewest number of packs that a mother of 4 children can buy if she wants to divide up the pieces equally among her children?
       (a) 1   **(b)** 3   **(c)** 12   **(d)** 2   **(e)** 6

    25. A bag of candy has 15 pieces. What is the fewest number of packs that a teacher of a class of 40 students can buy and be able to divide up the pieces equally among his pupils?
       (a) 40   **(b)** 120   **(c)** 20   **(d)** 8   **(e)** 5

    26. A necklace has 84 beads. There are 70 beads in a pack. How many packs should you buy so that you can make the fewest number of necklaces without having any beads left over?
       (a) 70   **(b)** 6   **(c)** 84   **(d)** 420   **(e)** 50

    27. A room is 45 feet by 60 feet. Carpeting comes in square pieces. What are the dimensions of the longest (equal sized) *squares* that can be bought so that they cover the room without cutting any pieces?
       (a) 5 ft by 5 ft   **(b)** 10 ft by 10 ft   **(c)** 6 ft by 6 ft
       (d) 15 ft by 15 ft   **(e)** 12 ft by 12 ft

SECTION G. Translating a Statement into a Mathematical Expression

An online store sells pens for $2 a pen plus $13 for shipping. The total charge for 30 pens is
$$(30)(\$2) + \$13 = \$73.$$
If n represents the number of pens purchased, then the charge for n pens is
$$2n + 13.$$

EXAMPLE 1: Translate each statement into a mathematical expression, where n represents the number.

    a. A certain number increased by 25.
    b. A certain number decreased by 45.
    c. Seven times the number.
    d. Twice the number decreased by 9.

SOLUTION:

    a. $n + 25$
    b. $n - 45$
    c. $7n$
    d. $2n - 9$

A **variable** is a symbol that represents an unknown quantity. In the above example the variable is the letter n.

The expression $7n$, is the product of variable n and **coefficient** 7. The coefficient is the number that is before the variable.

The expression $7x - 3y + 10$ has three **terms**: $7x$, $3y$, and 10. Terms are separated by addition or subtraction. The last term, 10, is called the constant term since it does not contain a variable. The number 7 is the coefficient of x, and -3 is the coefficient of y.

Two terms are called **like terms** if they have the same variable component. For example, $3x$ and $5x$ are like terms.

We can add all the like terms within an expression as follows,
$$5x + 3x = 8x \quad \text{To combine like terms, add the coefficients.}$$

33

EXAMPLE 2: Simplify the expression by combining like terms.
$$3x - 5y + 2x + 4y$$

SOLUTION: $3x - 5y + 2x + 4y =$
$3x + 2x - 5y + 4y =$
$5x \quad - \quad y \quad = 5x - y$

EXAMPLE 3: Simplify the expression by removing parenthesis and combining like terms.
$$3 - 2(4 - x) + 5x$$

SOLUTION: $3 - 2(4 - x) + 5x =$
$3 - 8 + 2x + 5x = -5 + 7x$

EXAMPLE 4: Let x represent the first of three numbers. The second number is 3 times the first, and the third number is one less than 2 times the first. Write expressions for the three numbers and for the sum of the three numbers.

SOLUTION:
The first number is x.
The second number is $3x$.
The third number is $2x - 1$.
The sum is $x + 3x + 2x - 1 = 6x - 1$.

Note: The coefficient of term x is 1.

We evaluate a mathematical expression when we assign a value to the variable in the expression. For example, if $n = 3$, then $7n + 2 = 7(3) + 2 = 21 + 2 = 23$.

EXAMPLE 5: Evaluate the mathematical expressions for $x = -3$.
    a. $4 + 2x$
    b. $5 - 7x$

SOLUTION: a. $4 + 2x = 4 + 2(-3)$
            $= 4 + (-6)$
            $= -2$

     b. $5 - 7x = 5 - 7(-3)$
            $= 5 - (-21)$
            $= 5 + 21$
            $= 26$

Exercises: Translating a Statement into a Mathematical Expression

For 1 – 4, translate each statement into a mathematical expression, where n represents the number.

1. A certain number increased by 5.

2. A certain number decreased by 10.

3. Five times a number.

4. Twice the number increased by 3.

For 5 – 12, express the relationships as mathematical expressions.

5. m added to x.

6. s less 4.

7. x divided by 2.

8. 5 times c.

9. p increased by 10.

10. The sum of 9 and b.

11. The difference of 7 and y.

12. 5 decreased by x.

For 13 – 16, simplify by combining like terms.

13. $5 + 3x + 10$

14. $3 + 4x + 5 + 5x$

15. $5y + 3x + 2y - 2x$

16. $4 - 6n + 5m - 3n + 2$

For 17 – 18, simplify by removing parenthesis and combining like terms.

17. $4 + 3(2 + x) + 9x$

18. $7 - 5(2 - 2x) + 3$

19. Let x represent the first of three numbers. The second number is 6 less than 5 times the first. The third number is 8 more than 4 times the first. Express the sum of the three numbers in terms of x.

(a) $9x - 6$ (b) $2x + 2$ (c) $10x + 14$ (d) $4x + 8$ (e) $10x + 2$

20. On Monday a boy sold x raffle tickets for his school. On Tuesday he sold 4 more than 3 times the number he sold on Monday. On Wednesday he sold twice the number he sold on Monday. Express the number of tickets the boy sold on the three days in terms of x.

(a) $3x + 8$ (b) $8x + 1$ (c) $3x - 2$ (d) $6x + 4$ (e) $7x + 2$

21. A boy sold $3x - 2$ newspapers one week, and $4 - 7x$ the next week. Express the difference between the number of papers sold on the first week and the number sold on the second week, in terms of x.

(a) $-x + 6$ (b) $7x - 9$ (c) $10x - 6$ (d) $-4x - 6$

For 22 – 25, evaluate the expressions for $x = -2$.

22. $7x - 5$ **23.** $3 - x$

24. $3(x - 4)$ **25.** $5(x - 4)$

36

SECTION H. Solving Applied Problems

EXAMPLE 1: A man buys a table and 6 chairs for $500. Each chair costs $50. How much did he pay for the table?

SOLUTION:
The cost of all the chairs is (6)($50) = $300.
Therefore, the cost of the table is $500 – $300 = $200.

EXAMPLE 2: The Smith children's favorite movie on DVD lasts 92 minutes. The family went on an 8 hour trip and the children watch the full DVD as many times as could. They only stopped watching when there was no longer enough time to repeat the entire movie. For how many minutes were they *not* watching the DVD?

SOLUTION:
There are (60)(8) = 480 minutes in 8 hours.
Divide 480 ÷ 92 = 5 Remainder 20.
Therefore, they did not watch the DVD for 20 minutes of the 8 hours.

EXAMPLE 3: A new stock was issued at an initial price of $10. The day it was issued a broker checked the stock price at the end of each hour for 7 hours. The price fluctuated by the following dollar amounts: +2, +3, +1, –1, –2, –3, and +2. What was the stock worth at the end of the day?

SOLUTION:
We combine all the price changes.
$2 + 3 + 1 + (-1) + (-2) + (-3) + 2 =$
$2 + 3 + 1 + 2 + (-1) + (-2) + (-3) =$ rearrange, separating the positives and negatives
$\quad 8 \quad + \quad (-6) \quad = 2$
Therefore, at the end of the day the stock was worth $10 + $2 = $12.

EXAMPLE 4: On opening day in a new shopping center, each 10th customer received a gift certificate, and each 15th customer received a pen set. Which customers received both a gift certificate and a pen set?

SOLUTION: The least common multiple of 10 and 15 is 30. The multiples of 30 are multiples of both 10 and 15. So the 30th, 60th, 90th, 120th, etc. customers received both gifts.

37

Exercises: Solving Applied Problems

1. A boy bought 5 tennis balls for $2 each, and 7 baseballs. The total was $31 dollars. How much did he pay for each baseball?
   (a) $3   (b) $2   (c) $21   (d) $4   (e) $5

2. A man bought a fridge for $500. He gave a deposit of $50 and he paid $20 when the fridge arrived. He paid the rest in 10 equal monthly installments. How much was each installment?
   (a) $45   (b) $30   (c) $50   (d) $43   (e) $25

3. Sue's table is 8 feet long and 4 feet wide. What is the area of the table top in square inches?
   (a) 384 sq in   (b) 4608 sq in   (c) 3200 sq in   (d) 3456 sq in   (e) 500 sq in

4. If 12 small tiles cover a medium sized tile, and 2 medium sized tiles cover a big tile, then how many small tiles cover 10 big tiles?
   (a) 36   (b) 120   (c) 48   (d) 24   (e) 240

5. Lew needs $3000 to buy a used car. He saved $53 each week for one full year. How much more does he need to save?
   (a) $250   (b) $244   (c) $1864   (d) $636   (e) $754

6. A math course has seven sections taught by seven different instructors. The final exam for the course consists of 65 questions. The teachers decide that they will each make up the same number of questions, and any questions that are left over will be chosen from the textbook. How many questions are chosen from the textbook?
   (a) 10   (b) 5   (c) 3   (d) 2   (e) 7

7. How many 8 inch square tiles can fit in 100 square feet?
   (a) 225   (b) 1500   (c) 180   (d) 1800   (e) 800

8. Latoya's garden is 8 yards by 12 yards. She buys grass seed that covers 1000 sq. ft. Did she buy more than enough or not enough? (She was over or under by how many square feet?)
   (a) under by 321 sq ft   (b) over by 543 sq ft   (c) over by 712 sq ft
   (d) under by 152 sq ft   (e) over by 136 sq ft

38

9. An online spice store charges $4 for a bottle of pepper, $7 for a bottle of saffron and $5 for a bottle of nutmeg. There is an $8 shipping and handling fee, and a $15 "small order charge" for orders that are under $25 dollars (not including shipping). Mariel orders two bottles of pepper, one bottle of saffron and one bottle of nutmeg. What is she charged for her order?

    (a) $43   **(b)** $35   **(c)** $45   **(d)** $65   **(e)** $38

10. The flight from New York to Toronto takes 87 minutes. If the plane takes off at 3:45 PM, what time does it land?

    (a) 5:55 PM   **(b)** 5:47 PM   **(c)** 5:12 PM   **(d)** 5:27 PM   **(e)** 4:57 PM

11. Karen is buying a tablecloth. Her table is 8ft. long. The store has a tablecloth that she likes that is 104 inches in length. How many inches of tablecloth will hang over the front on the table? Assume that the table is centered in the tablecloth.

    (a) 20   **(b)** 5   **(c)** 2   **(d)** 4   **(e)** 8

12. On a single day the temperature in Juneau, Alaska dropped from a high of $-12°F$ to a low of $-50°F$. Find the difference between high and low temperatures.

    (a) 38   **(b)** 68   **(c)** -38 **(d)** -62 **(e)** -45

13. Mark wrote a check for $50. He mistakenly adds this amount to his balance instead of subtracting it. As a result his balance is how much more than it should be?

    (a) $25   **(b)** $100   **(c)** $-$50   **(d)** $-$100   **(e)** $50

14. On five runs a football running back gained 5, 3, 0, -3, and -2. What was his total gain?

    (a) 13   **(b)** 3   **(c)** 10   **(d)** -3 **(e)** 5

15. On Monday morning Ken invested $100 in a volatile stock. He recorded his gains and or losses at the close of each day for five days. These were +$5, $-$2, $-$3, +$2, and $-$6. What was the value of his stock at the end of the fifth day?

    (a) $25   **(b)** $102   **(c)** $99   **(d)** $104   **(e)** $96

16. On a cold night in February, the temperature at 12:00 midnight was 23°F. Beginning at 1:00 AM the temperature changes recorded each hour were -2, -4, -1, -1, 0, and $+2$ degrees. What was the temperature at 6:00 AM?

    (a) $-6°$ **(b)** 17°   **(c)** 13°   **(d)** $-3°$ **(e)** $-2°$

17. A number is divisible by 63 and 42. Which number is not necessarily a divisor of the number?

(a) 6 **(b)** 12 **(c)** 18 **(d)** 14 **(e)** 21

18. A tile is 4 inches by 6 inches. What are the fewest number of tiles needed to form a square?

(a) 36 **(b)** 6 **(c)** 4 **(d)** 24 **(e)** 9

19. Let x represent the first of three numbers. The second number is x decreased by 2. The third number is three times x subtracted from 5. Express the sum of the three numbers in terms of x.

(a) $2x + 3$ **(b)** $6x - 7$ **(c)** $3x + 7$ **(d)** $-x + 3$ **(e)** $4x - 7$

20. A web hosting company charges a basic \$11 per month for 10 megabytes of server space and \$2 more for every megabyte over the initial 10. If x represents a number of megabytes, where x is greater than 10, write an expression for the monthly fee for x megabytes of server space.

(a) $11 + (x - 10)2$ **(b)** $10x + 11$ **(c)** $21x - 10$
(d) $(2x - 10) + 11$ **(e)** $11 + (10 - x)2$

21. A baker bakes an average of x loaves of bread each morning. On Monday he baked 20 loaves less than his average. On Tuesday he baked twice his average decreased by 50, and on Wednesday he baked 30 loaves more than his average. Express the total number of loaves baked on these three days in terms of x.

(a) $3x - 90$ **(b)** $4x - 40$ **(c)** $3x - 70$ **(d)** $2x - 80$ **(e)** $4x - 70$

22. A rectangle is $x + 3$ units by $2x - 3$ units. A square is $5x$ units by $5x$ units. What is the difference between the perimeter of the rectangle and the perimeter of the square?

(a) $5x$ **(b)** $14x$ **(c)** $-5x$ **(d)** $10x$ **(e)** $-14x$

23. The expression $(-4)(5) + \dfrac{3(2-5)}{(5-2)+6}$ is equal to _____.

(a) 23 **(b)** -12 **(c)** -21 **(d)** 21 **(e)** 12

Chapter 2

Fractions

A. Understanding Fractions: Proper Fractions, Improper Fractions and Mixed Numbers

A **fraction** is used to represent the number of equal parts of a whole. The fraction consists of three parts: the **numerator** (the top number), a slash or a fraction bar and the **denominator** (the bottom number). The denominator tells us how many equal parts the whole is divided into and the numerator tells us how many of these equal parts are being taken or talked about.

For example, if the horizontal bar below is divided into eight equal parts. The three shaded parts represent *part* of the whole and are represented by the fraction 3/8 or $\frac{3}{8}$.

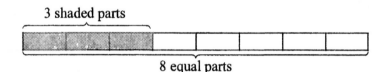

3 shaded parts

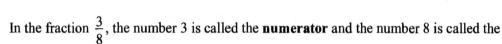

8 equal parts

In the fraction $\frac{3}{8}$, the number 3 is called the **numerator** and the number 8 is called the **denominator**.

fraction bar $\rightarrow \dfrac{3}{8} \begin{array}{l} \rightarrow \text{The } \textit{numerator} \text{ specifies how many of these parts are shaded.} \\ \rightarrow \text{The } \textit{denominator} \text{ specifies the total number of equal parts.} \end{array}$

When you say "$\frac{3}{4}$ of a loaf of bread has been eaten," what you are indicating is that the loaf of bread was cut into four equal pieces and three of the four parts have been eaten.

41

EXAMPLE 1: Use a fraction to represent the shaded part of the object.

SOLUTION:

    (a) Three out of four parts are shaded, or $\dfrac{3}{4}$.

    (b) Eight out of nine parts are shaded, or $\dfrac{8}{9}$.

    (c) Five out of five parts are shaded, or $\dfrac{5}{5} = 1$.

Note that in Example 1c, when 5 out of 5 parts are shaded, we have 1 whole amount. This illustrates a division fact which states that any nonzero number divided by itself is 1.

In addition to it representing a part of a whole, a fraction may represent a ratio of two numbers (see chapter 5), and may also mean division, in which case the numerator is the dividend and the denominator is the divisor. Thus $\dfrac{3}{8}$ may mean $3 \div 8$ or $8\overline{)3}$.

Since a fraction may mean division, the division rule involving zero applies to fractions. Namely: the value of a fraction is zero if the numerator is zero and the denominator is nonzero. If the denominator is zero, the fraction is undefined. Division by zero is not allowed.

We summarize as follows.

1. Any nonzero number divided by itself is 1.

$$8 \div 8 = \frac{8}{8} = 1$$

2. We can never divide by zero.

$$15 \div 0 = \frac{15}{0} \text{ is not allowed, i.e. division by zero is } \textbf{\textit{undefined}}.$$

3. If zero is divided by any number except zero, the result is always zero.

$$0 \div 3 = \frac{0}{3} = 0 \quad \text{In other words, } \textit{any fraction with 0 in the numerator}$$
$$\textit{and a nonzero denominator equals 0.}$$

EXAMPLE 2: Divide, if possible.

    (a) $\dfrac{4}{4}$ **(b)** $\dfrac{123}{0}$ **(c)** $\dfrac{0}{43}$

SOLUTION:

    (a) $\dfrac{4}{4} = 1$

    (b) $\dfrac{123}{0}$ Division by 0 is undefined.

    (c) $\dfrac{0}{43} = 0$ Any fraction with 0 in the numerator and a nonzero denominator equals 0.

EXAMPLE 3: The approximate number of inches of rain that falls during a selected period of one year in a certain city is given in the following table.

Time period	Number of inches in rainfall
January to March	23 inches
April to June	5 inches
July to September	4 inches
October to December	15 inches

    (a) What fractional part of the total yearly rainfall occurs from October to December?
    (b) What fractional part of the total yearly rainfall does *not* occur from July to September?

SOLUTION: First we must find the total rainfall for 1 year.

           23 in. + 15 in. + 4 in. + 5 in. = 47 in.

    (a) From October to December there were 15 inches of rain out of a total of 47 inches.

           $\dfrac{15}{47}$ Fractional part of rainfall that falls from October to December.

    (b) From July to September there were 4 inches of rain out of a total of 47 inches.

           47 in. − 4 in. = 43 in.

           $\dfrac{43}{47}$ Fractional part of rainfall that does *not* fall from July to September.

We have names for different kinds of fractions. A **proper fraction** is used to describe a quantity *less than 1*. If the numerator is *less* than the denominator, the fraction is a proper fraction. The fraction $\dfrac{3}{5}$ is a proper fraction.

An **improper fraction** is used to describe a quantity *greater* than or equal to 1. If the numerator is *greater than or equal to* the denominator, the fraction is an improper fraction. The fraction $\dfrac{9}{4}$ is an improper fraction because the numerator is larger than the denominator. Since $\dfrac{5}{5}$ describes a quantity equal to 1, it is also an improper fraction.

A **mixed number** is the sum of a whole number greater than zero and a proper fraction, and is used to describe a quantity greater than 1. An improper fraction can be written as a mixed number.

The following chart will help you visualize the different fractions and their names.

Value	Illustration	Math Symbol	Name
Less than 1		$\frac{4}{5}$	proper fraction
Equal to 1		$\frac{5}{5}$	improper fraction
Greater than 1		$\frac{5}{4}$ $1\frac{1}{4}$	improper fraction or mixed number

The last figure can also be represented by 1 whole added to $\frac{1}{4}$ of a whole or $1+\frac{1}{4}$. This is written $1\frac{1}{4}$ (the addition symbol is omitted). $1\frac{1}{4}$ is called a mixed number. Thus the improper fraction $\frac{5}{4}$ is equivalent to the mixed number $1\frac{1}{4}$.

EXAMPLE 4: Identify each as a proper fraction, an improper fraction, or a mixed number.

(a) $\frac{10}{9}$ (b) $\frac{7}{11}$ (c) $17\frac{3}{4}$ (d) $\frac{13}{13}$

SOLUTION:

(a) $\frac{10}{9}$ improper fraction   The numerator is larger than the denominator.

(b) $\frac{7}{11}$ proper fraction   The numerator is less than the denominator.

(c) $17\frac{3}{4}$ mixed number   A whole number is added to a proper fraction.

(d) $\frac{13}{13}$ improper fraction   The numerator is equal to the denominator.

B. Expressing an Improper Fraction as a Mixed Number and a Mixed Number as an Improper Fraction

To change an improper fraction such as $\frac{14}{5}$ into a mixed number, we could first think in terms of boards cut into fifths (i.e. each board is cut into 5 parts), as in the figure below. Here we have 2 whole boards (accounting for 10 parts) and 4 parts of another board which was also cut into 5 parts. We see that $\frac{14}{5} = 2\frac{4}{5}$ since 2 whole boards and $\frac{4}{5}$ of a board are shaded.

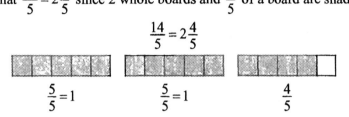

Numerically, we need to know how many times 5 goes into 14 (how many complete or whole boards we can get). We get 2 with 4 left over, i.e. we need 2 whole boards and $\frac{4}{5}$ of another. We can think of $\frac{14}{5}$ as 14 fractions $\frac{1}{5}$ lined up in the following way.

$$\underbrace{\frac{1}{5} \cdot \frac{1}{5} \cdot \frac{1}{5} \cdot \frac{1}{5} \cdot \frac{1}{5}}_{\frac{5}{5} = 1} \quad \underbrace{\frac{1}{5} \cdot \frac{1}{5} \cdot \frac{1}{5} \cdot \frac{1}{5} \cdot \frac{1}{5}}_{\frac{5}{5} = 1} \quad \underbrace{\frac{1}{5} \cdot \frac{1}{5} \cdot \frac{1}{5} \cdot \frac{1}{5}}_{\frac{4}{5}}$$

Thus $\frac{14}{5} = 2\frac{4}{5}$.

We summarize this procedure as follows.

To change an improper fraction into a mixed number, divide the numerator by the denominator. The quotient is the whole number part of the mixed number. The remainder is the new numerator of the fraction part and the denominator is the same as the improper fraction.

Thus the improper fraction
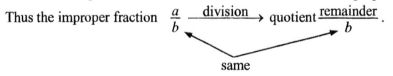

EXAMPLE 1: Write $\frac{19}{7}$ as a mixed number.

SOLUTION: The answer is in the form: quotient$\frac{\text{remainder}}{7}$.

Divide 19 by 7 to get a quotient of 2 and a remainder of 5. Thus $\frac{19}{7} = 2\frac{5}{7}$.

To change a mixed number to an improper fraction, we reverse our steps. In the mixed number $3\frac{3}{8}$, the 3 and the 8 tell us that there were 3 whole boards that were each divided up into 8 parts giving us $3 \times 8 = 24$ parts. The 3 in $\frac{3}{8}$ tells us that there were 3 additional parts. Then altogether there are $3 \times 8 + 3 = 27$ parts and the denominator 8 is the denominator of the improper fraction. Thus $3\frac{3}{8} = \frac{3 \times 8 + 3}{8} = \frac{27}{8}$

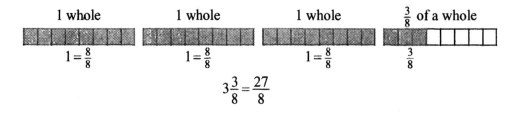

$$3\frac{3}{8} = \frac{27}{8}$$

We can summarize the procedure as follows:
To change a mixed number into an improper fraction, multiply the denominator by the whole number and add the numerator to this product. The resulting value is the numerator of the improper fraction. The denominator in the mixed number becomes the denominator of the improper fraction. Thus

$$\text{mixed number} \longrightarrow \text{improper fraction}$$
$$W\frac{N}{D} = \frac{W \cdot D + N}{D}$$

where W represents the whole number, N represents the numerator and D represents the denominator.

EXAMPLE 2: Change $6\frac{2}{3}$ to an improper fraction.

SOLUTION: Improper fraction: $\frac{W \cdot D + N}{D}$, where W is 6, D is 3 and N is 2.

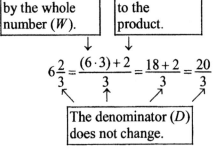

| Multiply the denominator (D) by the whole number (W). | Add the numerator (N) to the product. |

$$6\frac{2}{3} = \frac{(6 \cdot 3) + 2}{3} = \frac{18 + 2}{3} = \frac{20}{3}$$

The denominator (D) does not change.

Exercises: Understanding Fractions and Converting Improper Fractions to Mixed Numbers and Mixed Numbers to Improper Fractions

For 1 – 5, divide if possible.

1. $\dfrac{0}{7}$

2. $\dfrac{9}{0}$

3. $\dfrac{9}{9}$

4. $\dfrac{86}{86}$

5. $\dfrac{2}{0}$

For 6 – 13, choose the correct answer.

6. A baseball player had 7 base hits in 15 times at bat. Write the fraction that describes the number of times the player had a base hit.

    (a) $\dfrac{7}{15}$ (b) $\dfrac{15}{7}$ (c) $\dfrac{7}{22}$ (d) $\dfrac{15}{22}$ (e) not given

7. There are 57 women and 37 men in the hospital cafeteria. What fractional part of the customers in the hospital cafeteria consists of men?

    (a) $\dfrac{57}{37}$ (b) $\dfrac{37}{94}$ (c) $\dfrac{37}{57}$ (d) $\dfrac{57}{94}$ (e) not given

8. There are 26 dancers in the dance production class at a high school. Nine of the dancers are juniors. Write the fraction that describes the dancers who are not juniors.

    (a) $\dfrac{9}{26}$ (b) $\dfrac{26}{9}$ (c) $\dfrac{17}{26}$ (d) $\dfrac{9}{17}$ (e) not given

47

9. An archer hit the target 3 times out of 11 shots. Write the fraction that describes the number of times the archer hit the target.

   (a) $\frac{8}{11}$ (b) $\frac{11}{8}$ (c) $\frac{11}{3}$ (d) $\frac{3}{11}$ (e) not given

10. There are 87 men and 63 women working for a small corporation. What fractional part of the employees consists of men only?

    (a) $\frac{87}{63}$ (b) $\frac{63}{87}$ (c) $\frac{63}{150}$ (d) $\frac{150}{63}$ (e) not given

11. At a salad bar, there are 29 different items to choose from. Eleven of the choices contain pasta. Write the fraction that describes the choices that do not contain pasta.

    (a) $\frac{18}{29}$ (b) $\frac{29}{18}$ (c) $\frac{11}{29}$ (d) $\frac{11}{18}$ (e) not given

12. The city baseball team won 21 of the 32 games they played. What fractional part of the games did the team lose?

    (a) $\frac{21}{32}$ (b) $\frac{11}{32}$ (c) $\frac{11}{21}$ (d) $\frac{32}{11}$ (e) not given

13. A large candy bar is cut into 12 equal pieces. You ate 1 piece. What fraction of the candy bar did you eat?

    (a) $\frac{1}{3}$ (b) $\frac{1}{4}$ (c) $\frac{1}{6}$ (d) $\frac{1}{12}$ (e) not given

14. A pizza pie is cut into 8 equal pieces. You ate 3 pieces. What fraction of the pizza did you eat?

    (a) $\frac{1}{8}$ (b) $\frac{1}{4}$ (c) $\frac{3}{8}$ (d) $\frac{3}{4}$ (e) not given

15. Jim painted $\frac{3}{8}$ of his room. What fractional part of his room did he not paint?

    (a) $\frac{5}{8}$ (b) $\frac{4}{8}$ (c) $\frac{3}{8}$ (d) $\frac{2}{8}$ (e) not given

For 16 – 18, choose the correct mixed number which is the same as the given improper fraction.

16. $\frac{41}{2}$

    (a) $20\frac{1}{2}$ **(b)** $21\frac{2}{4}$ **(c)** $2\frac{1}{2}$ **(d)** $10\frac{3}{4}$ **(e)** not given

17. $\frac{19}{4}$

    (a) $4\frac{9}{19}$ **(b)** $4\frac{3}{4}$ **(c)** $4\frac{9}{4}$ **(d)** $19\frac{1}{4}$ **(e)** not given

18. $\frac{79}{7}$

    (a) $11\frac{1}{7}$ **(b)** $11\frac{2}{7}$ **(c)** $7\frac{9}{7}$ **(d)** $9\frac{7}{11}$ **(e)** not given

For 19 – 22, choose the correct improper fraction which is equivalent to the given mixed number.

19. $8\frac{2}{7}$

    (a) $\frac{82}{7}$ **(b)** $\frac{58}{7}$ **(c)** $\frac{89}{7}$ **(d)** $\frac{10}{7}$ **(e)** not given

20. $33\frac{1}{3}$

    (a) $\frac{34}{3}$ **(b)** $\frac{100}{3}$ **(c)** $\frac{111}{3}$ **(d)** $\frac{3}{111}$ **(e)** not given

21. $24\frac{1}{4}$

    (a) $\frac{61}{4}$ **(b)** $\frac{25}{4}$ **(c)** $\frac{97}{4}$ **(d)** $\frac{24}{6}$ **(e)** not given

22. $5\frac{19}{20}$

    (a) $\frac{95}{20}$ **(b)** $\frac{519}{20}$ **(c)** $\frac{20}{119}$ **(d)** $\frac{119}{20}$ **(e)** not given

C. Equivalent Fractions; Reducing Fractions to Lowest Terms

There are many fractions that name the same quantity. For example, in the following illustration the two pieces of wood are the same length.

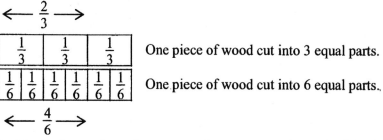

As you can see, 2 of the 3 pieces of wood represents the same quantity as 4 of the 6 pieces of wood. We say that $\frac{2}{3} = \frac{4}{6}$ and call these fractions **equivalent fraction**.

There are many other fractions that represent the same value as $\frac{2}{3}$ and thus are equivalent to $\frac{2}{3}$.

$$\frac{2}{3} = \frac{4}{6} = \frac{6}{9} = \frac{8}{12} \text{ are equivalent fractions.}$$

Equivalent fractions *look different* but have the same *value* because they represent the same quantity. Now, how can we find an equivalent fraction without using a picture or diagram? Observe the following pattern.

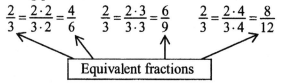

We can generalize: if we multiply both the numerator and denominator by the same nonzero number we get an equivalent fraction.

$$\frac{a}{b} = \frac{a \cdot c}{b \cdot c} \quad \text{where } b \text{ and } c \text{ are not 0.}$$

EXAMPLE 1:

  (a) Multiply the numerator and the denominator of $\frac{3}{4}$ by 2 to find an equivalent fraction.

  (b) Multiply the numerator and the denominator of $\frac{3}{4}$ by 3 to find an equivalent fraction.

SOLUTION: (a) $\frac{3}{4} = \frac{3 \cdot 2}{4 \cdot 2} = \frac{6}{8}$ (b) $\frac{3}{4} = \frac{3 \cdot 3}{4 \cdot 3} = \frac{9}{12}$

50

EXAMPLE 2: Write $\dfrac{3}{4}$ as an equivalent fraction with a denominator of 16.

SOLUTION:

$$\frac{3}{4} = \frac{}{16}$$

$$\frac{3 \cdot ?}{4 \cdot ?} = \frac{}{16} \quad \text{4 times what number equals 16? 4}$$

Since we must multiply the denominator by 4 to obtain 16 we must also multiply the numerator by 4.

$$\frac{3 \cdot 4}{4 \cdot 4} = \frac{12}{16}$$

A fraction is considered **reduced to lowest terms** and is said to be in **simplest form** when there is no common factor other than 1 in the numerator and the denominator.

reduced   not reduced

↓   ↓

$\dfrac{2}{3} = \dfrac{2 \cdot 1}{3 \cdot 1}$ $\dfrac{4}{6} = \dfrac{2 \cdot 2}{3 \cdot 2}$

no common factor   common factor

other than 1   of 2

The procedure for reducing a fraction to lowest terms is the reverse of finding the equivalent fraction.

$\dfrac{a}{b} = \dfrac{a \cdot c}{b \cdot c}$ $\dfrac{a \cdot \overset{1}{\cancel{c}}}{b \cdot \underset{1}{\cancel{c}}} = \dfrac{a}{b}$

$\longrightarrow$ $\longrightarrow$

creating equivalent   reducing to

fractions   lowest terms

Just as the value of a fraction is not changed if its numerator and denominator are both multiplied by the same nonzero number, so too the value of a fraction is not changed if its numerator and denominator are both divided by the same nonzero number.

Procedure to Reduce a Fraction to Lowest Terms

1. Write the numerator and denominator of the fraction as products of prime numbers. For example, $\frac{4}{6} = \frac{2 \cdot 2}{2 \cdot 3}$.

2. Any factor that appears in both the numerator and denominator is a common factor. Divide both numerator and denominator by every common factor.

$$\frac{2 \cdot 2}{2 \cdot 3} = \frac{2 \cdot \overset{1}{\cancel{2}}}{3 \cdot \underset{1}{\cancel{2}}} = \frac{2}{3}$$

3. Multiply the remaining factors.

The direction *simplify* or *reduce* means to reduce to lowest terms.

EXAMPLE 3: Simplify $\frac{15}{35}$

SOLUTION: Since the last digit of both 15 and 35 is 5, we know that both numbers are divisible by 5.

$$\frac{15}{35} = \frac{3 \cdot 5}{7 \cdot 5}$$ Step 1, write the numerator and denominator as products of prime numbers.

$$= \frac{3 \cdot \overset{1}{\cancel{5}}}{7 \cdot \underset{1}{\cancel{5}}}$$ Step 2, divide both numerator and denominator by the common factor 5.

$$= \frac{3 \cdot 1}{7 \cdot 1} = \frac{3}{7}$$ Step 3, multiply the remaining factors.

We can reduce a fraction with a negative number in either the numerator or the denominator by writing the negative sign in *front* of the fraction. Thus $\frac{-a}{b} = \frac{a}{-b} = -\frac{a}{b}$.

For example, $\frac{-3}{4} = \frac{3}{-4} = -\frac{3}{4}$.

When we reduce a fraction, we write the negative sign in front of the fraction. We do not include it as part of the prime factors.

EXAMPLE 4: Simplify $\dfrac{-72}{48}$

SOLUTION: If we recognize that the numerator and denominator have common factors that are *not* prime, we can use these factors to reduce the fraction.

$$\dfrac{-72}{48} = -\dfrac{72}{48} \qquad \text{Write the negative sign in front of the fraction.}$$

$$= -\dfrac{\overset{1}{\cancel{8}} \cdot 9}{\underset{1}{\cancel{8}} \cdot 6} \qquad \text{8 is a common factor of 72 and 48.}$$

$$-\dfrac{9}{6} = -\dfrac{3 \cdot \overset{1}{\cancel{3}}}{2 \cdot \underset{1}{\cancel{3}}} \qquad \text{Write the remaining factors as products of primes.}$$

$$= -\dfrac{3}{2} \qquad \text{Simplify.}$$

CAUTION: Students sometimes apply slashes incorrectly as follows.

| 3 is not a common factor. | $\dfrac{\cancel{3}+4}{\cancel{3}} = 4$ | THIS IS WRONG! |

$$\dfrac{3+4}{3} = \dfrac{7}{3} \qquad \text{THIS IS RIGHT!}$$

| Here, 3 is a common factor. | $\dfrac{\cancel{3} \cdot 4}{\cancel{3}} = \dfrac{4}{1} = 4$ | THIS IS RIGHT! |

We may not use slashes with addition or subtraction signs. We may use slashes only if we are dividing *common factors*.

EXAMPLE 5: Write $\dfrac{30}{12}$ in simplest form.

SOLUTION:

$$\dfrac{30}{12} = \dfrac{\overset{1}{\cancel{2}} \cdot \overset{1}{\cancel{3}} \cdot 5}{\underset{1}{\cancel{2}} \cdot 2 \cdot \underset{1}{\cancel{3}}} = \dfrac{5}{2} = 2\dfrac{1}{2}$$

$$\text{or} \quad \dfrac{30}{12} = \dfrac{\overset{1}{\cancel{6}} \cdot 5}{\underset{1}{\cancel{6}} \cdot 2} = \dfrac{5}{2} = 2\dfrac{1}{2}$$

EXAMPLE 6: Simplify $\dfrac{16}{24}$.

SOLUTION:

$$\frac{16}{24}=\frac{\overset{1}{\cancel{2}}\cdot\overset{1}{\cancel{2}}\cdot\overset{1}{\cancel{2}}\cdot2}{\underset{1}{\cancel{2}}\cdot\underset{1}{\cancel{2}}\cdot\underset{1}{\cancel{2}}\cdot3}=\frac{2}{3}$$

$$\text{or }\frac{16}{24}=\frac{\overset{1}{\cancel{8}}\cdot2}{\underset{1}{\cancel{8}}\cdot3}=\frac{2}{3}$$

Exercises: Equivalent Fractions; Reducing Fractions to Lowest Terms

For 1 – 6, find an equivalent fraction with the given denominator.

1. $\dfrac{3}{7} = \dfrac{?}{49}$

2. $\dfrac{3}{7} = \dfrac{?}{49}$

3. $\dfrac{3}{4} = \dfrac{?}{20}$

4. $\dfrac{7}{8} = \dfrac{?}{40}$

5. $\dfrac{9}{13} = \dfrac{?}{39}$

6. $\dfrac{8}{11} = \dfrac{?}{44}$

7. $\dfrac{35}{40} = \dfrac{?}{80}$

8. $\dfrac{45}{50} = \dfrac{?}{100}$

For 9 – 24, simplify each fraction.

9. $\dfrac{15}{25}$

10. $\dfrac{14}{21}$

11. $\dfrac{12}{16}$

12. $\dfrac{24}{30}$

13. $\dfrac{30}{60}$

14. $\dfrac{12}{32}$

15. $\dfrac{24}{28}$

16. $\dfrac{18}{27}$

17. $\dfrac{49}{35}$

18. $\dfrac{81}{72}$

19. $\dfrac{75}{60}$

20. $\dfrac{46}{23}$

21. $\dfrac{-24}{36}$

22. $\dfrac{-35}{40}$

23. $\dfrac{-42}{48}$

24. $\dfrac{-40}{50}$

D. Multiplying and Dividing Fractions and Mixed Numbers

Consider a bar divided into 5 equal parts, 4 of which are shaded, i.e. $\frac{4}{5}$ of the bar is shaded.

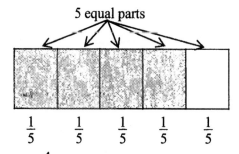

Now shade $\frac{2}{3}$ of the $\frac{4}{5}$ that is already shaded, i.e. divide the shaded portion into 3 equal parts and shade 2 of these 3 parts.

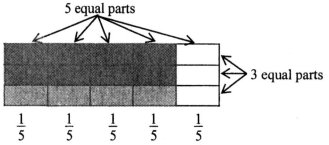

Notice that the original bar is actually divided into 15 equal parts of which 8 are now shaded.

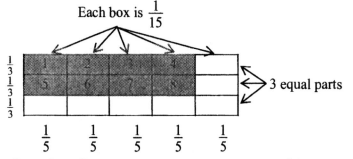

We thus see that $\frac{2}{3}$ of $\frac{4}{5}$ is $\frac{8}{15}$. Therefore, to find two-thirds of $\frac{4}{5}$, we multiply $\frac{2}{3} \times \frac{4}{5} = \frac{8}{15}$. The product of two fractions is the product of the numerators divided by the product of the denominators. The product $\frac{2}{3} \times \frac{4}{5}$ is read "$\frac{2}{3}$ times $\frac{4}{5}$" or "$\frac{2}{3}$ of $\frac{4}{5}$". So when we say "take a fraction of something" the word "of" indicates multiplication.

EXAMPLE 1: Multiply $\frac{1}{3} \times \frac{1}{2}$.

SOLUTION: $\frac{1}{3} \times \frac{1}{2} = \frac{1 \cdot 1}{3 \cdot 2} = \frac{1}{6}$

EXAMPLE 2: What is $\frac{2}{3}$ of $\frac{2}{5}$?

SOLUTION: $\frac{2}{3} \times \frac{2}{5} = \frac{2 \cdot 2}{3 \cdot 5} = \frac{4}{15}$

Even before actually multiplying numerators and denominators, we should try to factor out common factors of the numerator and denominator. After the common factors are eliminated, multiply the numbers remaining in the numerator and multiply the numbers remaining in the denominator.

EXAMPLE 3: Multiply $\frac{9}{10} \times \frac{2}{3}$

SOLUTION: $\frac{9}{10} \times \frac{2}{3} = \frac{9 \cdot 2}{10 \cdot 3} = \frac{\cancel{3} \cdot 3 \cdot \cancel{2}}{\cancel{2} \cdot 5 \cdot \cancel{3}} = \frac{3}{5}$

EXAMPLE 4: Find $\frac{18}{20} \times \frac{5}{21}$

SOLUTION: $\frac{18}{20} \times \frac{5}{21} = \frac{18 \cdot 5}{20 \cdot 21} = \frac{3 \cdot \cancel{3} \cdot \cancel{2} \cdot \cancel{5}}{\cancel{5} \cdot \cancel{2} \cdot 2 \cdot \cancel{3} \cdot 7} = \frac{3}{14}$

EXAMPLE 5: What is $\frac{4}{15}$ of $\frac{5}{28}$?

SOLUTION: $\frac{4}{15} \times \frac{5}{28} = \frac{\cancel{2} \cdot \cancel{2} \cdot \cancel{5}}{3 \cdot \cancel{5} \cdot \cancel{2} \cdot \cancel{2} \cdot 7} = \frac{1}{21}$

57

To multiply mixed numbers, first change the mixed number to improper fraction and then multiply the fractions. Whole numbers should be written as a fraction with a denominator of 1.

EXAMPLE 6: Multiply $4\frac{5}{6} \times \frac{12}{17}$.

SOLUTION: $4\frac{5}{6} \times \frac{12}{17} = \frac{29}{6} \times \frac{12}{17}$ Convert $4\frac{5}{6} = \frac{29}{6}$

$$= \frac{29 \cdot \overset{1}{\cancel{6}} \cdot 2}{\underset{1}{\cancel{6}} \cdot 17} = \frac{58}{17} = 3\frac{7}{17}$$

EXAMPLE 7: Find $\frac{5}{8} \times 6$.

SOLUTION: $\frac{5}{8} \times \frac{6}{1} = \frac{5 \cdot 6}{8 \cdot 1} = \frac{5 \cdot 3 \cdot \overset{1}{\cancel{2}}}{4 \cdot \underset{1}{\cancel{2}} \cdot 1} = \frac{15}{4} = 3\frac{3}{4}$

EXAMPLE 8: Find $5\frac{2}{3} \times 4\frac{1}{2}$.

SOLUTION: $5\frac{2}{3} \times 4\frac{1}{2} = \frac{17}{3} \times \frac{9}{2}$ Convert $5\frac{2}{3} = \frac{17}{3}$ and $4\frac{1}{2} = \frac{9}{2}$

$$= \frac{17 \cdot 9}{3 \cdot 2} = \frac{17 \cdot \overset{1}{\cancel{3}} \cdot 3}{\underset{1}{\cancel{3}} \cdot 2} = \frac{51}{2} = 25\frac{1}{2}$$

EXAMPLE 9: Multiply $-\frac{2}{16} \cdot \frac{8}{20}$.

SOLUTION: When multiplying signed fractions, first determine the sign of the product and then multiply.

$$-\frac{2}{16} \cdot \frac{8}{20} = -\frac{2 \cdot 8}{16 \cdot 20} \qquad \text{The product of a negative and a positive is negative.}$$

$$= -\frac{\overset{1}{\cancel{2}} \cdot \overset{1}{\cancel{8}}}{\underset{1}{\cancel{8}} \cdot \underset{1}{\cancel{2}} \cdot 5 \cdot 4} = -\frac{1}{20}$$

As we know division is the reverse of multiplication. a divided by b can be written as

$$a \div b, a/b, \text{ or } \frac{a}{b}.$$

We note that $6 \times \frac{1}{3}$ is the same as $\frac{6}{3}$. Thus multiplying 6 by $\frac{1}{3}$ is the same as dividing 6 by 3. We call 3 and $\frac{1}{3}$ reciprocals. The reciprocal of a fraction is the fraction with the numerator and denominator interchanged. To find the reciprocal of a fraction we interchange (invert) the numerator and denominator. Thus dividing a by b is the same as multiplying a by the reciprocal of b. Therefore to divide a fraction by a fraction we simply invert the denominator fraction and multiply it by the numerator.

EXAMPLE 10: Divide $\frac{5}{8} \div \frac{4}{9}$.

SOLUTION: $\frac{5}{8} \div \frac{4}{9} = \frac{5}{8} \times \frac{9}{4}$ The reciprocal of $\frac{4}{9}$ is $\frac{9}{4}$.

$\qquad\qquad = \frac{5 \cdot 9}{8 \cdot 4} = \frac{45}{32} = 1\frac{13}{32}$

Mixed numbers should first be converted to improper fractions and whole numbers should be written as fractions with denominators of 1.

EXAMPLE 11: Divide:

    (a) $8 \div \frac{3}{4}$ (b) $\frac{4}{9} \div 6$ (c) $\frac{3}{8} \div 3\frac{1}{4}$ (d) $2\frac{3}{4} \div 1\frac{4}{9}$

SOLUTION:

    (a) $8 \div \frac{3}{4} = \frac{8}{1} \div \frac{3}{4}$ Write 8 as $\frac{8}{1}$

$\qquad\qquad = \frac{8}{1} \times \frac{4}{3}$ Invert and multiply

$\qquad\qquad = \frac{8 \cdot 4}{1 \cdot 3} = \frac{32}{3} = 10\frac{2}{3}$

    (b) $\frac{4}{9} \div 6 = \frac{4}{9} \div \frac{6}{1} = \frac{4}{9} \times \frac{1}{6} = \frac{4 \cdot 1}{9 \cdot 6} = \frac{2 \cdot \overset{1}{\cancel{2}} \cdot 1}{3 \cdot 3 \cdot 3 \cdot \underset{1}{\cancel{2}}} = \frac{2}{27}$

    (c) $\frac{3}{8} \div 3\frac{1}{4} = \frac{3}{8} \div \frac{13}{4}$ Convert $3\frac{1}{4}$ to improper fraction $\frac{13}{4}$

$\qquad\qquad = \frac{3}{8} \times \frac{4}{13} = \frac{3 \cdot 4}{8 \cdot 13} = \frac{3 \cdot \overset{1}{\cancel{4}}}{2 \cdot \underset{1}{\cancel{4}} \cdot 13} = \frac{3}{26}$

    (d) $2\frac{3}{4} \div 1\frac{4}{9} = \frac{11}{4} \div \frac{13}{9}$ Convert mixed numbers to improper fractions.

$\qquad\qquad = \frac{11}{4} \times \frac{9}{13} = \frac{99}{52} = 1\frac{47}{52}$

Exercises: Multiplying and Dividing Fractions and Mixed Numbers

For 1 – 3, find each of the following.

1. $\frac{1}{3}$ of $\frac{1}{5}$ 2. $\frac{1}{5}$ of $\frac{1}{7}$ 3. $\frac{5}{21}$ of $\frac{7}{8}$

For 4 – 21, multiply and simplify your answer.

4. $\frac{7}{12} \cdot \frac{8}{28}$ 5. $\frac{7}{12} \cdot \frac{8}{28}$

6. $\frac{3}{42} \cdot \frac{6}{15}$ 7. $\frac{-3}{8} \cdot \left(\frac{32}{-6}\right)$

8. $\frac{-5}{2} \cdot \left(\frac{2}{-30}\right)$ 9. $\frac{16}{11} \cdot \left(\frac{-18}{36}\right)$

10. $\frac{6}{35} \cdot 5$ 11. $\frac{2}{21} \cdot 15$

12. $-14 \cdot \frac{1}{28}$ 13. $\frac{-2}{63} \cdot \left(\frac{-14}{18}\right)$

14. $\frac{-8}{20} \cdot \left(\frac{-25}{32}\right)$ 15. $\frac{1}{5} \cdot 25$

16. $2\frac{5}{8} \cdot 4\frac{2}{3}$ 17. $2 \cdot 5\frac{1}{8}$

18. $3\frac{1}{9} \cdot 3\frac{3}{8}$ 19. $\frac{2}{3} \cdot 9$

20. $21\frac{5}{8} \cdot 28$ 21. $3\frac{1}{9} \cdot 3\frac{3}{8}$

For 22 – 39, divide and simplify your answer.

22. $\frac{6}{14} \div \frac{3}{8}$ 23. $\frac{8}{12} \div \frac{5}{6}$

24. $\frac{7}{24} \div \frac{9}{8}$ 25. $\frac{9}{28} \div \frac{4}{7}$

26. $\frac{-1}{12} \div \frac{3}{4}$ 27. $\frac{-1}{15} \div \frac{2}{3}$

28. $\frac{-7}{24} \div \left(\frac{9}{-8}\right)$ 29. $\frac{-9}{28} \div \left(\frac{4}{-7}\right)$

30. $15 \div \frac{3}{7}$ 31. $18 \div \frac{2}{3}$

32. $\frac{7}{22} \div 14$ 33. $\frac{8}{26} \div 16$

34. $4\frac{1}{5} \div 21$ 35. $6\frac{9}{16} \div 1\frac{3}{32}$

36. $9\frac{7}{12} \div 1\frac{1}{12}$ 37. $\frac{5}{6} \div 20$

38. $6\frac{1}{2} \div \frac{1}{2}$ 39. $\frac{5}{12} \div 2\frac{2}{5}$

40. $4 \times \frac{1}{4} = ?$

   (a) $\frac{1}{16}$ (b) $\frac{4}{16}$ (c) $\frac{5}{16}$ (d) $4\frac{1}{4}$ (e) not given

61

E. Solving Application Problems

EXAMPLE 1:   If a $2\frac{1}{2}$ gallon gasoline container is $\frac{1}{3}$ full, how many gallons of gasoline does it contain?

SOLUTION: We must compute $\frac{1}{3}$ of $2\frac{1}{2}$:

$$\frac{1}{3}\cdot\frac{5}{2} \qquad\qquad 2\frac{1}{2}=\frac{5}{2}$$

$$\frac{1\cdot 5}{3\cdot 2}=\frac{5}{6}\text{ gallons}$$

EXAMPLE 2:   A recipe calls for $\frac{3}{4}$ of a cup of sugar. How much sugar should be used to make $\frac{1}{2}$ of the recipe?

SOLUTION: We must compute $\frac{1}{2}$ of $\frac{3}{4}$.

$$\frac{1}{2}\cdot\frac{3}{4}$$

$$\frac{1\cdot 3}{2\cdot 4}=\frac{3}{8}\text{ of a cup of sugar should be used}$$

EXAMPLE 3:   Stan is putting on molding on his walls. Each piece of molding is 45 inches long. The walls he is working on has a total length of 20 yards. What fraction of the walls to be worked on would equal one piece of molding?

SOLUTION: We look for the fraction of the length of a piece of molding (numerator) and the length of the walls (denominator), both expressed in the same units. Thus,

length of molding 45 in.

length of walls   20 yards $\times$ 36 in = 720 in

the required fraction is $\dfrac{45}{720}=\dfrac{\overset{1}{\cancel{5}}\cdot\overset{1}{\cancel{9}}}{\underset{1}{\cancel{9}}\cdot 8\cdot\underset{1}{\cancel{5}}\cdot 2}=\dfrac{1}{16}$

EXAMPLE 4: We are told that $\frac{1}{5}$ of all people in a certain city have blue eyes. About how many blue-eyed people would you expect in a group of 5000 people from this city?

SOLUTION: We must compute $\frac{1}{5}$ of 5000.

$$\frac{1}{5} \cdot \frac{5000}{1}$$

$$\frac{1 \cdot 5000}{5 \cdot 1} = \frac{1 \cdot \overset{1}{\cancel{5}} \cdot 1000}{\underset{1}{\cancel{5}} \cdot 1} = 1000 \text{ blue-eyed people are expected in a group of 5000.}$$

EXAMPLE 5: Jim borrowed \$4800 from his father. For Jim's birthday present, his father said he would reduce his debt by \$2000 or by $\frac{3}{8}$ of the loan, which ever is greater. How much does Jim owe his father after Jim's birthday?

SOLUTION: $\frac{3}{8}$ of 4800

$$\frac{3}{8} \cdot \frac{4800}{1} = \frac{3 \cdot \overset{1}{\cancel{8}} \cdot 600}{\underset{1}{\cancel{8}} \cdot 1} = 1800 \text{ which is less than 2000.}$$

So Jim's father will reduce Jim's loan by \$2000. So now Jim owes \$4800 – \$2000 = \$2800.

EXAMPLE 6: A can of food contains $2\frac{3}{4}$ servings with 100 calories in each serving. How many calories are there in $\frac{1}{5}$ of a can of this food?

SOLUTION: First we get the total calories in the entire can: $2\frac{3}{4} \cdot 100$.

$$\frac{11}{4} \cdot \frac{100}{1} \qquad\qquad 2\frac{3}{4} = \frac{11}{4}$$

$$\frac{11 \cdot \overset{1}{\cancel{4}} \cdot 25}{\underset{1}{\cancel{4}} \cdot 1} = 275 \text{ calories for a whole can}$$

So in $\frac{1}{5}$ can, there are $\frac{1}{5} \cdot \frac{275}{1} = \frac{1 \cdot \overset{1}{\cancel{5}} \cdot 55}{\underset{1}{\cancel{5}}} = 55$ calories in $\frac{1}{5}$ can.

EXAMPLE 7: Two-thirds of 2 hours 24 minutes is equal to how many hours?

SOLUTION: Convert 2 hours 24 minutes into minutes:

2 hours 24 minutes $= 2 \times 60 + 24 = 144$ minutes

$$\frac{2}{3} \cdot 144 = \frac{2 \cdot \cancel{3} \cdot 48}{\cancel{3}} = 96 \text{ minutes}$$

$$96 \text{ minutes} = \frac{96}{60} \text{ hours; we simplify}$$

$$\frac{96}{60} = \frac{\cancel{6} \cdot \cancel{2} \cdot 8}{\cancel{6} \cdot \cancel{2} \cdot 5} = \frac{8}{5} = 1\frac{3}{5} \text{ hours}$$

EXAMPLE 8: In a classroom, $\frac{1}{4}$ of the students are female. Two-thirds of the male students have brown eyes. What fraction of the class is composed of brown-eyed male students?

SOLUTION: Since $\frac{1}{4}$ of the students are female, $\frac{3}{4}$ of the students are male. To find the fraction of brown-eyed male students, we must find $\frac{2}{3}$ of $\frac{3}{4}$:

$$\frac{2}{3} \cdot \frac{3}{4} = \frac{\cancel{2} \cdot \cancel{3}}{\cancel{3} \cdot \cancel{2} \cdot 2} = \frac{1}{2}, \text{ i.e. } \frac{1}{2} \text{ of the class are brown-eyed males.}$$

EXAMPLE 9: How many empty sugar bowls, each $\frac{2}{3}$ cup capacity, can be filled using all the sugar in a bag that contains 14 cups of sugar?

SOLUTION: To find the number of bowls that can be filled, we find out how many $\frac{2}{3}$ cups there are in 14 cups, i.e. we divide 14 by $\frac{2}{3}$.

$$\frac{14}{\frac{2}{3}} = \frac{14}{1} \cdot \frac{3}{2} \qquad \text{We invert and multiply.}$$

$$= \frac{\cancel{2} \cdot 7 \cdot 3}{1 \cdot \cancel{2}} = 21 \text{ bowls}$$

EXAMPLE 10: If three-eights of a long pipe is 30 feet, how long is the pipe?

SOLUTION: If $\frac{3}{8}$ of a pipe is 30 feet, then $\frac{1}{8}$ of the pipe is $\frac{1}{3}$ of 30 = 10 feet long. If $\frac{1}{8}$ of the pipe is 10 feet then $\frac{8}{8}$ = the entire pipe is 8×10 feet $= 80$ feet.

In general, if we know what is a fraction of the whole, as in example 10, we know that $\frac{3}{8}$ of the whole pipe is 30 feet. Then to find the length of the whole pipe we simply divide the 30 feet by the fractional part $\frac{3}{8}$. Thus

$$30 \div \frac{3}{8} = \frac{30}{1} \cdot \frac{8}{3} = \frac{\cancel{3} \cdot 10 \cdot 8}{1 \cdot \cancel{3}} = 80 \text{ feet}$$

As another example if we know that $\frac{2}{3}$ of a pound of candy costs $\$2\frac{1}{2}$. Then one whole pound costs

$$2\frac{1}{2} \div \frac{2}{3} = \frac{5}{2} \div \frac{2}{3} = \frac{5}{2} \cdot \frac{3}{2} = \frac{15}{4} = \$3\frac{3}{4}$$

EXAMPLE 11: If Carol used $9\frac{1}{4}$ gallons of gasoline to travel 111 miles, how many miles can she travel using 1 gallon?

SOLUTION: We need to compute the number of miles per gallon by dividing the distance 111 miles by the number of gallons it took to travel that distance. Thus:

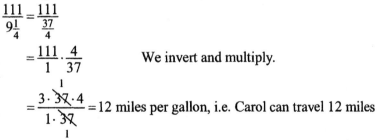

$$\frac{111}{9\frac{1}{4}} = \frac{111}{\frac{37}{4}}$$

$$= \frac{111}{1} \cdot \frac{4}{37} \qquad \text{We invert and multiply.}$$

$$= \frac{3 \cdot \cancel{37} \cdot 4}{1 \cdot \cancel{37}} = 12 \text{ miles per gallon, i.e. Carol can travel 12 miles}$$

using 1 gallon of gasoline.

EXAMPLE 12: A 29-foot pipe is to be cut into $\frac{3}{4}$ foot pieces for use in a construction project. What is the length of the remaining piece after as many $1\frac{3}{4}$ foot pieces as possible have been cut?

SOLUTION: To get the number of $1\frac{3}{4}$ foot pieces that can be obtained from a 29 foot pipe, divide 29 by $1\frac{3}{4}$.

$$\frac{29}{1\frac{3}{4}} = \frac{29}{\frac{7}{4}}$$
$$= \frac{29}{1} \cdot \frac{4}{7} \qquad \text{We invert and multiply.}$$
$$= \frac{116}{7} = 16\frac{4}{7}$$

i.e. from a 29 foot pipe we can cut 16 complete $1\frac{3}{4}$ foot pipes with $\frac{4}{7}$ of a $1\frac{3}{4}$ foot pipe leftover. Thus we compute $\frac{4}{7}$ of $1\frac{3}{4}$ foot to get the length of the 29 foot pipe that is leftover.

$$\frac{4}{7} \cdot \frac{7}{4} = \frac{4 \cdot 7}{7 \cdot 4} = 1 \text{ foot}, \text{ i.e. the length of the piece remaining is 1 foot.}$$

Or we can look at this situation in the following way. Since there can be 16 complete $1\frac{3}{4}$ foot pipes, the total length of these 16 pipes is $16 \times 1\frac{3}{4} = \frac{16}{1} \times \frac{7}{4} = \frac{4 \cdot 4 \cdot 7}{1 \cdot 4} = 28$. But we started with a 29 foot long pipe. Thus there remains 29 − 28 = 1 foot.

EXAMPLE 13: If 24 ounces of water fills $\frac{1}{4}$ of a jar. How many ounces of water will fill $\frac{1}{3}$ of the same jar?

SOLUTION: Since 24 ounces fills $\frac{1}{4}$ of a jar, 4 times 24 will fill the entire jar, i.e. the entire jar, therefore has $4 \times 24 = 96$ oz capacity. So that $\frac{1}{3}$ of the jar is $\frac{1}{3} \cdot 96 = 32$ oz, i.e. 32 oz of water fills $\frac{1}{3}$ of the jar.

EXAMPLE 14: A large container of milk weighs 13 pounds. After $\frac{1}{3}$ of milk is spilled out, the container and the remaining milk weighs $9\frac{1}{2}$ pounds, how much does the empty container weigh?

SOLUTION: $\frac{1}{3}$ of the milk accounts for the decrease in weight from 13 to $9\frac{1}{2}$ pounds, i.e. $\frac{1}{3}$ of the milk weighs $3\frac{1}{2}$ pounds. So that the milk weighs $3 \times 3\frac{1}{2} = \frac{3}{1} \times \frac{7}{2} = \frac{3 \cdot 7}{1 \cdot 2} = \frac{21}{2} = 10\frac{1}{2}$ lbs. So the container itself weighs $13 - 10\frac{1}{2} = 2\frac{1}{2}$ lbs.

EXAMPLE 15: John and Al run a 6-mile race. John runs at a constant rate of $8\frac{1}{2}$ minutes per mile and Al runs at a constant rate of 8 minutes per mile. How many minutes more will Al have to run to complete the race after John finishes the race.

SOLUTION: John will take $8\frac{1}{2} \times 6 = \frac{17}{2} \times \frac{6}{1} = \frac{17 \cdot 3 \cdot \overset{1}{\cancel{2}}}{\underset{1}{\cancel{2}} \cdot 1} = 51$ minutes to finish the race. Al will take $8 \times 6 = 48$ minutes to finish the race. Thus Al will need to run $51 - 48 = 3$ minutes after John finishes the race.

Exercises: Solving Application Problems

1. A person can walk $4\frac{1}{2}$ miles in 1 hour. How far can the person walk in $\frac{1}{3}$ hour?

   (a) $1\frac{1}{2}$ mi   **(b)** 2 mi   **(c)** 3 mi   **(d)** 4 mi   **(e)** $2\frac{1}{2}$ mi

2. A compact car travels 42 miles on each gallon of gasoline. How many miles can the car travel on $9\frac{1}{2}$ gallons of gasoline?

   (a) 300 mi   **(b)** 450 mi   **(c)** 399 mi   **(d)** 100 mi   **(e)** not given

3. A board is $6\frac{1}{4}$ feet long. One fifth of the board is cut off. What is the length of the piece cut off?

   (a) $2\frac{1}{2}$ **(b)** $2\frac{1}{2}$ **(c)** $3\frac{1}{4}$ **(d)** $1\frac{1}{4}$ **(e)** not given

4. A cook is using a recipe that calls for $3\frac{1}{3}$ cups of flour. The cook wants to triple the recipe. How much flour should the cook use?

   (a) 9 cups   **(b)** 11 cups   **(c)** 10 cups   **(d)** 20 cups   **(e)** not given

5. The F-1 engine in the first stage of the Saturn 5 rocket burns 214,000 gallons of propellant in 1 minute. The first stage burns $2\frac{1}{2}$ minutes before burnout. How much propellant is used before burnout?

   (a) 500,000 gallons   **(b)** 250,000 gallons   **(c)** 530,000 gallons
   (d) 535,000 gallons   **(e)** not given

6. The stock of a high technology company is selling for $\$63\frac{3}{8}$. Find the cost of 48 shares of the stock.

   (a) $4200   **(b)** $3282   **(c)** $2800   **(d)** $3100   **(e)** not given

7. A soft drink company budgets $\frac{1}{10}$ of its income each month for advertising. In July, the company had an income of $24,500. What is the amount budgeted for advertising in July?
   (a) $2400   **(b)** $2450   **(c)** $2500   **(d)** $2550   **(e)** not given

8. The parents of the school choir members are making robes for the choir. Each robe requires $2\frac{5}{8}$ yards of material at a cost of $8 per yard. Find the total cost of 24 choir robes.
   (a) $502   **(b)** $503   **(c)** $504   **(d)** $552   **(e)** $560

9. A college spends $\frac{5}{8}$ of its monthly income on employee salaries. During one month the college had an income of $712,000. How much of the monthly income remained after the employee's salaries were paid?
   (a) $610,000   **(b)** $250,000   **(c)** $576,000   **(d)** $376,000   **(e)** $267,000

10. Individual cereal boxes contain $\frac{3}{4}$ ounce of cereal. How many boxes can be filled with 400 ounces of cereal?
    (a) 600   **(b)** 700   **(c)** 750   **(d)** 800   **(e)** not given

11. A box contains 25 ounces of cereal. How many $1\frac{1}{4}$–ounce portions can be served from this box?
    (a) 14   **(b)** 16   **(c)** 18   **(d)** 20   **(e)** not given

12. A sales executive used $18\frac{2}{5}$ gallons of gas on a 368-mile trip. How many miles can the sales executive travel on 1 gallon of gas?
    (a) 10 mi   **(b)** 20 mi   **(c)** 30 mi   **(d)** 40 mi   **(e)** 50 mi

13. A commuter plane used $53\frac{3}{4}$ gallons of fuel on a 5-hour flight. How many gallons were used each hour?
    (a) $10\frac{3}{4}$ **(b)** $11\frac{3}{8}$ **(c)** $12\frac{1}{4}$ **(d)** $11\frac{5}{8}$ **(e)** not given

14. A $\frac{5}{8}$ carat diamond was purchased for $1200. What would a similar diamond weighing 1 carat cost?
(a) $1820   (b) $1920   (c) $1760   (d) $1850   (e) $2050

15. A utility stock is offered for 4\frac{3}{4}$ per share. How many shares can you buy for $304?
(a) 55   (b) 54   (c) 64   (d) 70   (e) not given

16. A new play opened, and 1200 people attended the opening at the concert hall. The hall was $\frac{2}{3}$ full. What is the capacity of the concert hall?
(a) 600   (b) 800   (c) 1200   (d) 1800   (e) 1900

17. One-tenth of a shipment of $11\frac{1}{9}$ pounds of grapes was spoiled. How many $\frac{1}{2}$–pound packages of unspoiled grapes can be packaged from this shipment?
(a) 20   (b) 25   (c) 30   (d) 15   (e) 40

18. A 16-foot board is cut into pieces $3\frac{1}{2}$ feet long for a bookcase. What is the length of the piece remaining after as many shelves as possible have been cut?
(a) 1 ft   (b) $1\frac{1}{2}$ ft   (c) 2 ft   (d) $2\frac{1}{2}$ ft   (e) $2\frac{3}{4}$ ft

19. A recipe requires $\frac{3}{4}$ cup of sugar. Which of the following calculators gives the number of cups of sugar that should be used to make $\frac{1}{3}$ of the recipe?
(a) $\frac{1}{3} \times \frac{3}{4}$ (b) $\frac{1}{3} \div \frac{3}{4}$ (c) $\frac{3}{4} \div \frac{1}{3}$ (d) $\frac{3}{4} + \frac{1}{3}$ (e) $\frac{3}{4} - \frac{1}{3}$

20. Gail has $\frac{3}{16}$ of her monthly income placed in a savings account. What is her monthly income if $180 is placed in her savings account?
(a) $760   (b) $860   (c) $960   (d) $460   (e) $360

21. How many bottles, each with $\frac{3}{4}$ cup capacity, can be filled using all milk in a 12-cup container?

    (a) 12   **(b)** 14   **(c)** 16   **(d)** 18   **(e)** 20

22. If a $4\frac{1}{2}$-gallon container is $\frac{3}{8}$ full, the number of gallons it contains is

    (a) less than $\frac{1}{2}$ **(b)** more than $\frac{1}{2}$ but less than 1

    (c) more than 1 but less than $1\frac{1}{2}$ **(d)** more than $1\frac{1}{2}$ but less than 2

    (e) more than 2

23. If 18 ounces of milk fills $\frac{1}{3}$ of a container, how many ounces will fill $\frac{1}{2}$ of this container?

    (a) 25 oz   **(b)** 27 oz   **(c)** 29 oz   **(d)** 31 oz   **(e)** 33 oz

24. Consider two sticks A and B. Stick A is 9 inches long and stick B is 4 yards long. What fraction of stick B would be equivalent to the length of stick A?

    (a) $\frac{1}{6}$ **(b)** $\frac{1}{10}$ **(c)** $\frac{1}{16}$ **(d)** $\frac{1}{32}$ **(e)** not given

25. A jar of juice weighs 12 pounds. $\frac{1}{4}$ of the juice spills out. The jar of juice now weighs $9\frac{1}{2}$ pounds. How much does the empty jar weigh?

    (a) 1 lb   **(b)** $1\frac{1}{2}$ lb   **(c)** 2 lb   **(d)** $2\frac{1}{2}$ lb   **(e)** not given

26. Carol and Susan will have a 10-mile bike race. Carol rides at a constant rate of 3 minutes per mile, and Susan rides at a constant rate of 4 minutes per mile. What distance, in miles, will Susan have to travel at the moment Carol is finishing the race?

    (a) 1 mile   **(b)** 2 miles   **(c)** $3\frac{1}{3}$ miles   **(d)** $3\frac{1}{2}$ miles   **(e)** not given

F. Adding and Subtracting Fractions and Mixed Numbers

Consider the following situation. You have a long board divided into 8 parts. Shade 3 parts, i.e. $\frac{3}{8}$ of the entire board. Then add an additional 2 parts, i.e. $\frac{2}{8}$ of the board and shade them.

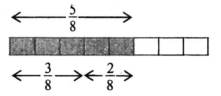

We note that $\frac{5}{8}$ of the board is shaded. We can represent the situation in symbols as follows.

$$\frac{3}{8} + \frac{2}{8} = \frac{3+2}{8} = \frac{5}{8}$$

From the figure, we can also see subtraction. If we start with 5 parts, $\left(\text{i.e. } \frac{5}{8}\right)$, $\frac{5}{8} - \frac{3}{8}$, we have 2

parts left $\left(\text{i.e. } \frac{2}{8}\right)$. In symbolic form:

$$\frac{5}{8} - \frac{3}{8} = \frac{2}{8} = \frac{1}{4}$$

We note:
1. The fraction that we added or subtracted had the same denominator (common denominator)
2. We added or subtracted numerators only.
3. The denominator in the result is the same as the common denominator.

In general, the procedure to add or subtract fractions is:
1. The fraction must have the same denominator.
2. Add or subtract the numerator only.
3. Keep the common denominator in the answer.
4. Write the sum or difference in simplest form.

EXAMPLE 1: Compute.

    (a) $\frac{3}{7} + \frac{2}{7}$ **(b)** $\frac{3}{8} + \frac{5}{8}$ **(c)** $\frac{7}{9} - \frac{2}{9}$ **(d)** $\frac{9}{9} - \frac{8}{9}$

SOLUTION:

    (a) $\frac{3}{7} + \frac{2}{7} = \frac{3+2}{7} = \frac{5}{7}$

    (b) $\frac{3}{8} + \frac{5}{8} = \frac{3+5}{8} = \frac{8}{8} = 1$

    (c) $\frac{7}{9} - \frac{2}{9} = \frac{7-2}{9} = \frac{5}{9}$

    (d) $\frac{9}{9} - \frac{8}{9} = \frac{9-8}{9} = \frac{1}{9}$

To add or subtract fractions with unlike (different) denominators, we first must convert the fractions into equivalent fractions with the same denominator. This common denominator will be the least common multiple (LCM) of the denominators (see chapter 1). The LCM of the denominators is usually called the least common denominator (LCD).

EXAMPLE 2: Find the LCD of the fractions.

     (a) $\dfrac{1}{2}, \dfrac{1}{3}$ **(b)** $\dfrac{3}{8}, \dfrac{7}{12}$ **(c)** $\dfrac{5}{8}, \dfrac{7}{9}$

SOLUTION:
     (a) LCM of 2 and 3 is 6.
     (b) LCM of 8 and 12

$$8 = 2 \cdot 2 \cdot 2$$
$$12 = 2 \cdot 2 \cdot 3$$
$$\text{LCM} = 2 \cdot 2 \cdot 2 \cdot 3 = 24$$

     (c) LCM of 8 and 9

$$8 = 2 \cdot 2 \cdot 2$$
$$9 = 3 \cdot 3$$
$$\text{LCM} = 2 \cdot 2 \cdot 2 \cdot 3 \cdot 3 = 72$$

EXAMPLE 3: Compute.

     (a) $\dfrac{1}{2} + \dfrac{1}{3}$ **(b)** $\dfrac{3}{8} + \dfrac{7}{12}$ **(c)** $\dfrac{7}{9} - \dfrac{5}{8}$

SOLUTION:
     (a) Convert to equivalent fraction with LCD 6 (see Example 2a)

$$\frac{1 \cdot 3}{2 \cdot 3} = \frac{3}{6}; \quad \frac{1 \cdot 2}{3 \cdot 2} = \frac{2}{6}$$
$$\frac{3}{6} + \frac{2}{6} = \frac{3+2}{6} = \frac{5}{6}$$

     (b) LCD 24 (see Example 2b)

$$\frac{3 \cdot 3}{8 \cdot 3} = \frac{9}{24}; \quad \frac{7 \cdot 2}{12 \cdot 2} = \frac{14}{24}$$
$$\frac{9}{24} + \frac{14}{24} = \frac{9+14}{24} = \frac{23}{24}$$

     (c) LCD 72 (see Example 2c)

$$\frac{7 \cdot 8}{9 \cdot 8} = \frac{56}{72}; \quad \frac{5 \cdot 9}{8 \cdot 9} = \frac{45}{72}$$
$$\frac{56}{72} - \frac{45}{72} = \frac{56-45}{72} = \frac{11}{72}$$

To add or subtract mixed numbers, simply convert the mixed number to an improper fraction and proceed as with adding and subtracting fractions.

EXAMPLE 4:   Compute.

(a) $3 + \dfrac{5}{8}$ (b) $5\dfrac{4}{9} + 4\dfrac{14}{15}$ (c) $11\dfrac{1}{2} + 5\dfrac{2}{3} + 2\dfrac{3}{8}$ (d) $5\dfrac{1}{3} - 2\dfrac{3}{4}$ (e) $12 - 7\dfrac{1}{11}$

SOLUTION:

(a) $3 + \dfrac{5}{8} = \dfrac{3}{1} + \dfrac{5}{8}$ Write as fraction

$\quad \dfrac{24}{8} + \dfrac{5}{8}$ LCD $= 8;\ \dfrac{3 \cdot 8}{1 \cdot 8} = \dfrac{24}{8}$

$\quad \dfrac{24 + 5}{8} = \dfrac{29}{8} = 3\dfrac{5}{8}$

(b) $5\dfrac{4}{9} + 4\dfrac{14}{15} = \dfrac{49}{9} + \dfrac{74}{15}$ Convert mixed numbers into improper fractions

$\quad \dfrac{245}{45} + \dfrac{222}{45}$ LCD of 9 and 15 is 45; $\dfrac{49 \cdot 5}{9 \cdot 5} = \dfrac{245}{45};\ \dfrac{74 \cdot 3}{15 \cdot 3} = \dfrac{222}{45}$

$\quad \dfrac{245 + 222}{45} = \dfrac{467}{45} = 10\dfrac{17}{45}$

(c) $11\dfrac{1}{2} + 5\dfrac{2}{3} + 2\dfrac{3}{8} = \dfrac{23}{2} + \dfrac{17}{3} + \dfrac{19}{8}$ Convert mixed numbers into improper fractions

$\quad \dfrac{276}{24} + \dfrac{136}{24} + \dfrac{57}{24}$ LCD of 2, 3 and 8 is 24; $\dfrac{23 \cdot 12}{2 \cdot 12} = \dfrac{276}{24};$

$\quad \dfrac{276 + 136 + 57}{24} = \dfrac{469}{24} = 19\dfrac{13}{24}$ $\dfrac{17 \cdot 8}{3 \cdot 8} = \dfrac{136}{24};\ \dfrac{19 \cdot 3}{8 \cdot 3} = \dfrac{57}{24}$

(d) $5\dfrac{1}{3} - 2\dfrac{3}{4} = \dfrac{16}{3} - \dfrac{11}{4}$ Convert to improper fractions

$\quad \dfrac{16 \cdot 4}{3 \cdot 4} - \dfrac{11 \cdot 3}{4 \cdot 3}$ LCD of 3 and 4 is 12

$\quad \dfrac{64}{12} - \dfrac{33}{12}$ Subtract

$\quad \dfrac{64 - 33}{12} = \dfrac{31}{12} = 2\dfrac{7}{12}$

(e) $12 - 7\dfrac{1}{11} = \dfrac{12}{1} - \dfrac{78}{11}$

$\quad \dfrac{12 \cdot 11}{1 \cdot 11} - \dfrac{78}{11}$

$\quad \dfrac{132}{11} - \dfrac{78}{11}$

$\quad \dfrac{132 - 78}{11} = \dfrac{54}{11} = 4\dfrac{10}{11}$

EXAMPLE 5: Add $\dfrac{5}{40}+\dfrac{3}{8}+\dfrac{1}{30}$.

SOLUTION: The LCD of 40, 8 and 30 is 120.

$$\frac{5}{40}=\frac{5\cdot 3}{40\cdot 3}=\frac{15}{120}; \quad \frac{3\cdot 15}{8\cdot 15}=\frac{45}{120}; \quad \frac{1\cdot 4}{30\cdot 4}=\frac{4}{120}$$

$$\frac{5}{40}+\frac{3}{8}+\frac{1}{30}=\frac{15}{120}+\frac{45}{120}+\frac{4}{120}=\frac{15+45+4}{120}$$

$$=\frac{64}{120}=\frac{\overset{1}{\cancel{8}}\cdot 8}{\underset{1}{\cancel{8}}\cdot 15}=\frac{8}{15}$$

EXAMPLE 6: Perform the indicated operation $\dfrac{5}{30}+\dfrac{3}{40}-\dfrac{1}{8}$.

SOLUTION: The LCD of 30, 40 and 8 is 120.

$$\frac{5\cdot 4}{30\cdot 4}=\frac{20}{120}; \quad \frac{3\cdot 3}{40\cdot 3}=\frac{9}{120}; \quad \frac{1\cdot 15}{8\cdot 15}=\frac{15}{120}$$

$$\frac{5}{30}+\frac{3}{40}-\frac{1}{8}=\frac{20}{120}+\frac{9}{120}-\frac{15}{120}=\frac{20+9-15}{120}$$

$$=\frac{14}{120}=\frac{7\cdot\overset{1}{\cancel{2}}}{60\cdot\underset{1}{\cancel{2}}}=\frac{7}{60}$$

75

Exercises: Adding and Subtracting Fractions and Mixed Numbers

For 1 – 40, perform the indicated operations. Simplify your answer.

1. $\dfrac{11}{15} - \dfrac{31}{45}$

2. $\dfrac{21}{12} - \dfrac{23}{24}$

3. $\dfrac{16}{24} - \dfrac{1}{6}$

4. $\dfrac{11}{28} - \dfrac{1}{7}$

5. $\dfrac{3}{8} + \dfrac{4}{7}$

6. $\dfrac{7}{4} + \dfrac{5}{9}$

7. $\dfrac{5}{6} + \dfrac{4}{5}$

8. $\dfrac{3}{7} + \dfrac{7}{2}$

9. $\dfrac{-3}{4} + \dfrac{1}{5}$

10. $\dfrac{-5}{6} + \dfrac{3}{4}$

11. $\dfrac{-2}{13} + \dfrac{7}{26}$

12. $\dfrac{-4}{15} + \dfrac{11}{30}$

13. $\dfrac{-3}{14} + \left(\dfrac{-9}{28}\right)$

14. $\dfrac{-4}{9} + \left(\dfrac{-11}{18}\right)$

15. $\dfrac{3}{16} + \left(\dfrac{-9}{20}\right)$

16. $\dfrac{3}{18} + \left(\dfrac{-5}{27}\right)$

17. $\dfrac{7}{10} - \dfrac{13}{100}$

18. $\dfrac{3}{10} - \dfrac{7}{100}$

19. $\dfrac{9}{8} - \dfrac{5}{32}$

20. $\dfrac{9}{50} - \dfrac{2}{25}$

21. $11\dfrac{2}{3} + 7\dfrac{1}{4}$

22. $22\dfrac{3}{5} + 16\dfrac{1}{10}$

23. $7\dfrac{4}{5} - 2\dfrac{1}{10}$

24. $6\dfrac{3}{8} - 2\dfrac{1}{16}$

25. $9\dfrac{2}{3} - 6\dfrac{1}{6}$

26. $15\dfrac{3}{4} - 13\dfrac{1}{6}$

27. $1\dfrac{1}{6} + \dfrac{3}{8}$

28. $1\dfrac{2}{3} + \dfrac{5}{18}$

29. $25\dfrac{2}{3} - 6\dfrac{1}{7}$

30. $45\dfrac{3}{8} - 26\dfrac{1}{16}$

31. $14\dfrac{7}{9} + 6\dfrac{1}{3}$

32. $13\dfrac{1}{2} + 7\dfrac{4}{5}$

33. $\dfrac{1}{3} - \dfrac{2}{9}$

　(a) $\dfrac{1}{9}$　　(b) $\dfrac{2}{9}$　　(c) $\dfrac{3}{9}$　　(d) $\dfrac{1}{27}$　　(e) $\dfrac{1}{12}$

34. $\dfrac{1}{5} - \dfrac{3}{20}$

　(a) $-\dfrac{2}{100}$　　(b) $-\dfrac{2}{25}$　　(c) $\dfrac{1}{15}$　　(d) $\dfrac{2}{20}$　　(e) $\dfrac{1}{20}$

35. $\dfrac{3}{8}+\dfrac{1}{4}$

    (a) $\dfrac{3}{32}$ **(b)** $\dfrac{4}{32}$ **(c)** $\dfrac{4}{12}$ **(d)** $\dfrac{5}{8}$ **(e)** not given

36. $\dfrac{1}{2}-\dfrac{1}{8}$

    (a) $\dfrac{2}{16}$ **(b)** $\dfrac{3}{16}$ **(c)** $\dfrac{3}{10}$ **(d)** $\dfrac{3}{8}$ **(e)** $\dfrac{2}{6}$

37. $\dfrac{3}{5}+\dfrac{3}{10}$

    (a) $\dfrac{6}{15}$ **(b)** $\dfrac{9}{150}$ **(c)** $\dfrac{9}{10}$ **(d)** $\dfrac{1}{5}$ **(e)** not given

38. $\dfrac{1}{12}+\dfrac{3}{14}+\dfrac{5}{21}$

    (a) $\dfrac{41}{84}$ **(b)** $\dfrac{9}{49}$ **(c)** $\dfrac{15}{28}$ **(d)** $\dfrac{17}{84}$ **(e)** not given

39. $\dfrac{1}{5}+\dfrac{2}{3}-\dfrac{13}{15}$

    (a) $-\dfrac{10}{7}$ **(b)** 0 **(c)** $\dfrac{26}{15}$ **(d)** $\dfrac{11}{15}$ **(e)** not given

40. $\dfrac{1}{3}+\dfrac{5}{12}-\dfrac{1}{6}$

    (a) $\dfrac{7}{12}$ **(b)** $\dfrac{3}{4}$ **(c)** $\dfrac{1}{4}$ **(d)** $\dfrac{11}{12}$ **(e)** not given

G. Solving Application Problems

EXAMPLE 1:   Stan and Bill are painting a room. If Stan has painted $\frac{1}{8}$ of the room and Bill has painted another $\frac{1}{4}$ of the room, what fraction of the room remains to be painted?

SOLUTION: To find the fractional part of the room that has yet to be painted: find the total amount of the room already painted $\left(\frac{1}{8}+\frac{1}{4}\right)$, then subtract the amount already done from 1, which represents the complete painted room.

$$\frac{1}{8}+\frac{1}{4}= \qquad\qquad \text{LCD of 8 and } 4 = 8$$

$$\frac{1}{8}+\frac{1\cdot2}{4\cdot2}=\frac{1}{8}+\frac{2}{8}$$

$$=\frac{1+2}{8}=\frac{3}{8}$$

Therefore, $1-\frac{3}{8}=\frac{8}{8}-\frac{3}{8}=\frac{8-3}{8}=\frac{5}{8}$ of the room has yet to be painted.

EXAMPLE 2:   In a recent survey, 38 out of every 50 customers at a major supermarket chose brand X, while 9 out of every 25 customers chose brand Y. The fraction who chose brand X is how much greater than the fraction who chose brand Y?

SOLUTION: We must subtract the brand Y fraction $\left(\frac{9}{25}\right)$ from the brand X fraction $\left(\frac{38}{50}\right)$.

$$\frac{38}{50}-\frac{9}{25}=\frac{38}{50}-\frac{9\cdot2}{25\cdot2}$$

$$=\frac{38}{50}-\frac{18}{50}=\frac{38-18}{50}$$

$$=\frac{20}{50}=\frac{2\cdot\overset{1}{\cancel{10}}}{5\cdot\underset{1}{\cancel{10}}}=\frac{2}{5}$$

EXAMPLE 3: What fraction added to the sum of $\frac{1}{8} + \frac{2}{3}$ will give a total of 1?

SOLUTION: We must find the sum of $\frac{1}{8} + \frac{2}{3}$ and then subtract the sum from 1.

$$\frac{1}{8} + \frac{2}{3} = \frac{1 \cdot 3}{8 \cdot 3} + \frac{2 \cdot 8}{3 \cdot 8} \qquad \text{LCD of 8 and 3 is 24}$$

$$= \frac{3}{24} + \frac{16}{24} = \frac{3+16}{24} = \frac{19}{24}$$

$$1 - \frac{19}{24} = \frac{24}{24} - \frac{19}{24}$$

$$= \frac{24-19}{24} = \frac{5}{24}$$

EXAMPLE 4: Jack and Jill take turns mowing the lawn. Jack mowed $\frac{3}{8}$ of the lawn on the first day and on the second day Jill mowed $\frac{1}{2}$ of what was left to be mowed. What fraction of the lawn remains to be mowed after the second day?

SOLUTION: At the end of the first day: $\frac{3}{8}$ mowed by Jack. So $1 - \frac{3}{8} = \frac{8}{8} - \frac{3}{8} = \frac{5}{8}$ of the lawn remained to be mowed. Jill then mows $\frac{1}{2}\left(\frac{5}{8}\right) = \frac{5}{16}$ of the lawn. So that at the end of the second day, $\frac{3}{8} + \frac{5}{16} = \frac{6}{16} + \frac{5}{16} = \frac{6+5}{16} = \frac{11}{16}$ of the lawn is mowed leaving $1 - \frac{11}{16} = \frac{16}{16} - \frac{11}{16} = \frac{16-11}{16} = \frac{5}{16}$ of the lawn yet to be mowed.

EXAMPLE 5: What number is $\frac{1}{3}$ the distance on the number line from $3\frac{1}{4}$ to $7\frac{3}{4}$ starting from $3\frac{1}{4}$?

SOLUTION: The distance from $3\frac{1}{4}$ to $7\frac{3}{4}$ is $7\frac{3}{4} - 3\frac{1}{4} = \frac{31}{4} - \frac{13}{4} = \frac{19}{4} = \frac{9}{2} = 4\frac{1}{2}$. $\frac{1}{3}$ of that distance is $\frac{1}{3} \cdot \frac{9}{2} = \frac{1 \cdot 9}{3 \cdot 2} = \frac{1 \cdot \overset{1}{\cancel{3}} \cdot 3}{\underset{1}{\cancel{3}} \cdot 2} = \frac{3}{2}$. Thus we must add $\frac{3}{2}$ to $3\frac{1}{4}$;

$$\frac{3}{2} + 3\frac{1}{4} = \frac{3 \cdot 2}{2 \cdot 2} + \frac{13}{4} = \frac{6}{4} + \frac{13}{4} = \frac{19}{4} = 4\frac{3}{4}.$$

EXAMPLE 6: A board 15 feet long is cut into 4 pieces: one piece was 2 feet $7\frac{1}{2}$ inches long, another piece was 3 feet 4 inches, a third piece was 4 feet $3\frac{1}{2}$ inches. The saw removes and additional $\frac{1}{3}$ inch at each cutting. What is the length, in feet, of the fourth piece?

SOLUTION: We must add the lengths of the 3 pieces and $\frac{1}{3}$ inch at each of the 3 cuttings and subtract that sum from 15 ft. It would be easier to work in units of inches so we first convert the lengths to inches: 2 feet $7\frac{1}{2}$ inches $= 2(12) + 7\frac{1}{2} = 31\frac{1}{2}$ inches

$$3 \text{ feet 4 inches} = 3(12) + 4 = 40 \text{ inches}$$

$$4 \text{ feet } 3\frac{1}{2} \text{ inches} = 4(12) + 3\frac{1}{2} = 51\frac{1}{2} \text{ inches}$$

$$3 \cdot \frac{1}{3} \text{ inches} = \qquad\qquad 1 \text{ inch; total length removed by the cuttings}$$

$$31\frac{1}{2} + 40 + 51\frac{1}{2} + 1 = \frac{63}{2} + 40 + \frac{103}{2} + 1$$
$$= \frac{63}{2} + \frac{40 \cdot 2}{1 \cdot 2} + \frac{103}{2} + \frac{1 \cdot 2}{1 \cdot 2}$$
$$= \frac{63}{2} + \frac{80}{2} + \frac{103}{2} + \frac{2}{2}$$
$$= \frac{63 + 80 + 103 + 2}{2}$$
$$= \frac{248}{2}$$
$$= 124 \text{ inches}$$

We now must subtract 124 inches from 15 feet: 15 feet $= 15 \times 12 = 180$ inches . Then the fourth piece is 180 in. $- 124$ in. $= 56$ in. or $\frac{56}{12} = 4\frac{8}{12}$ or 4 feet 8 inches or $4\frac{2}{3}$ feet.

EXAMPLE 7: John died and left an estate to be divided equally among his 5 children. His son Carl's portion of the estate was to be equally divided among his 6 children. How much of John's estate was inherited by 1 of Carl's children?

SOLUTION: John's son Carl inherited $\frac{1}{5}$ of the estate. This portion is now to be subdivided into 6 parts, i.e. each of Carl's children will inherit $\frac{1}{6}$ of $\frac{1}{5}$ of the estate. Thus each will inherit

$$\frac{1}{6} \cdot \frac{1}{5} = \frac{1 \cdot 1}{6 \cdot 5} = \frac{1}{30}.$$

EXAMPLE 8: A worker earns \$8 an hour. He worked 5, $3\frac{1}{2}$, $7\frac{1}{6}$, $5\frac{2}{3}$, and $7\frac{2}{3}$ hours during the last 5 days. Find his total wages for the 5 days.

SOLUTION: First we find the total number of hours he worked during the last 5 days and then multiply that sum by \$8. Thus:

$$5 + 3\frac{1}{2} + 7\frac{1}{6} + 5\frac{2}{3} + 7\frac{2}{3} = \frac{5}{1} + \frac{7}{2} + \frac{43}{6} + \frac{17}{3} + \frac{23}{3}$$

$$= \frac{5 \cdot 6}{1 \cdot 6} + \frac{7 \cdot 3}{2 \cdot 3} + \frac{43}{6} + \frac{17 \cdot 2}{3 \cdot 2} + \frac{23 \cdot 2}{3 \cdot 2}$$

$$= \frac{30}{6} + \frac{21}{6} + \frac{43}{6} + \frac{34}{6} + \frac{46}{6}$$

$$= \frac{30 + 21 + 43 + 34 + 46}{6}$$

$$= \frac{174}{6} = 29 \text{ hours}$$

and now multiplying by 8 yields

$$29 \times 8 = \$232$$

Exercises: Solving Application Problems

1. A $3\frac{2}{3}$ inch piece of board is cut from a $6\frac{7}{8}$ inch board. How much of the board is left?

    (a) $3\frac{1}{2}$ in.   **(b)** $4\frac{1}{4}$ in.   **(c)** $3\frac{15}{24}$ in.   **(d)** $4\frac{7}{24}$ in.   **(e)** not given

For 2 – 3, a cookie recipe calls for $\frac{3}{4}$ cup of granulated sugar and $\frac{7}{8}$ cup powdered sugar.

2. How many cups of sugar are in the recipe?

    (a) $\frac{1}{8}$ **(b)** $\frac{3}{8}$ **(c)** $\frac{7}{8}$ **(d)** $1\frac{5}{8}$ **(e)** not given

3. How many more cups of powdered sugar than granulated sugar are in the recipe?

    (a) $\frac{1}{8}$ **(b)** $\frac{3}{8}$ **(c)** $\frac{7}{8}$ **(d)** $1\frac{5}{8}$ **(e)** not given

For 4 – 6, a mother and her teenage son are painting their home. In one day, the mother completes $\frac{1}{3}$ of the job and the son completes $\frac{1}{4}$ of the job.

4. What was the fraction of the job completed?

    (a) $\frac{1}{3}$ **(b)** $\frac{1}{7}$ **(c)** $\frac{2}{7}$ **(d)** $\frac{7}{12}$ **(e)** not given

5. How must more of the job did the mother complete than the son?

    (a) $\frac{1}{7}$ **(b)** $\frac{2}{7}$ **(c)** $\frac{2}{3}$ **(d)** $\frac{1}{12}$ **(e)** not given

6. How much of the job remains to be done?

    (a) $\frac{5}{12}$ **(b)** $\frac{2}{3}$ **(c)** $\frac{6}{7}$ **(d)** $\frac{5}{7}$ **(e)** not given

7. A plumber has a pipe $15\frac{3}{4}$ feet long. He needs $\frac{1}{9}$ of the pipe for a repair job. What length must he cut off the pipe to get what he needs?

   (a) $14\frac{1}{2}$ ft.   (b) $1\frac{3}{4}$ ft.   (c) $10\frac{3}{4}$ ft.   (d) $1\frac{7}{9}$ ft.   (e) not given

8. What fraction added to the sum $\frac{1}{4}+\frac{3}{8}$ will get a total of 1?

   (a) $\frac{3}{8}$ (b) $\frac{1}{8}$ (c) $\frac{1}{4}$ (d) $\frac{1}{2}$ (e) $\frac{5}{8}$

9. Bill and Bob take turns painting a house. Bill painted $\frac{1}{4}$ of the house on Sunday. On Monday, Bob painted $\frac{3}{8}$ of the part of the house that was left unpainted by Bill. What fraction of the house remains to be painted after Monday?

   (a) $\frac{3}{8}$ (b) $\frac{5}{8}$ (c) $\frac{15}{32}$ (d) $\frac{13}{32}$ (e) not given

10. In a recent survey it was found that 3 out of every 10 people surveyed listened to station XYZ and 4 out of every 6 people surveyed listened to station ABC. The fraction who listen to station ABC is how much greater than the fraction who listen to station XYZ?

    (a) $\frac{11}{30}$ (b) $\frac{1}{4}$ (c) $\frac{2}{30}$ (d) $\frac{5}{8}$ (e) not given

11. Last year Bob bought some shares of stock in an oil company. This year Bob bought more shares of stock in the same oil company. He now has $1\frac{1}{2}$ times as many shares as he had at the beginning of this year. Today the shares are worth \$10 each and the total value for all his shares are \$3000. How many shares did he buy last year?

    (a) 100   (b) 200   (c) 300   (d) 50   (e) not given

12. A saw blade is $\frac{3}{16}$ inches thick and when it is used to cut a wood board, $\frac{3}{16}$ of an inch of the board is turned into sawdust. If we need to cut a board into 4 pieces by making 3 cuts with this saw blade, how much of the board, in inches, is turned into sawdust by the blade?

    (a) $\frac{3}{8}$ (b) $\frac{9}{16}$ (c) $\frac{3}{5}$ (d) $\frac{5}{8}$ (e) not given

84

For 13 – 14, a plumber has to cut 2 pieces from a pipe that is 25 feet long. If one piece is 5 feet $5\frac{1}{2}$ inches long and the other piece is 12 feet $4\frac{1}{4}$ inches long, how long of a pipe will remain

13. If we assume that no part of the pipe is removed because of the saw cuts?

    (a) 7 feet $2\frac{1}{4}$ inches   **(b)** 17 feet $9\frac{3}{4}$ inches   **(c)** 8 feet $9\frac{1}{2}$ inches

    (d) 18 feet $8\frac{1}{4}$ inches   **(e)** not given

14. If the 2 saw cuts eliminates $\frac{1}{8}$ of an inch for each cut of the pipe?

    (a) 6 feet $10\frac{1}{2}$ inches   **(b)** 7 feet 2 inches   **(c)** 17 feet $9\frac{1}{2}$ inches

    (d) 8 feet $9\frac{1}{4}$ inches   **(e)** not given

15. A family of four finds that $\frac{1}{3}$ of their income is spent on housing, $\frac{1}{10}$ is spent on transportation, and $\frac{1}{4}$ is spent on food. Find the total fractional amount of their income that is spent on food, housing and transportation combined.

    (a) $\frac{41}{60}$ **(b)** $\frac{15}{60}$ **(c)** $\frac{21}{60}$ **(d)** $\frac{3}{17}$ **(e)** not given

16. At the beginning of the year, the stock in a cellular phone company was selling for $\$26\frac{5}{8}$ per share. The price of the stock gained $\$14\frac{3}{4}$ per share during a 6-month period. Find the price of the stock at the end of 6 months.

    (a) $\$40\frac{5}{8}$ **(b)** $\$40\frac{7}{8}$ **(c)** $\$41\frac{3}{8}$ **(d)** $\$42\frac{1}{4}$ **(e)** not given

For 17 – 18, a 25-mile walk a thon for Breast Cancer has 3 checkpoints: A, B, and C. Checkpoint A is $3\frac{5}{8}$ miles from the starting point. Checkpoint B is $4\frac{2}{3}$ miles from the first checkpoint A.

17. How many miles is it from the starting point to checkpoint B?

   (a) $7\frac{7}{11}$ mi.   **(b)** $8\frac{7}{24}$ mi.   **(c)** $8\frac{1}{3}$ mi.   **(d)** $16\frac{17}{24}$ mi.   **(e)** $20\frac{1}{4}$ mi.

18. How many miles is it from the second checkpoint B to the finishing line?

   (a) $8\frac{7}{24}$ mi.   **(b)** $7\frac{7}{11}$ mi.   **(c)** $16\frac{17}{24}$ mi.   **(d)** $8\frac{1}{3}$ mi.   **(e)** $20\frac{1}{4}$ mi.

For 19 – 20, consider 2 points A and B on the number line. A is at $12\frac{1}{4}$ units from the origin O, and B is $22\frac{1}{2}$ units from O.

19. What number is halfway between A and B?

   (a) $10\frac{1}{4}$ **(b)** $17\frac{3}{8}$ **(c)** $5\frac{1}{8}$ **(d)** $18\frac{1}{2}$ **(e)** not given

20. What number is $\frac{1}{4}$ of the distance from A to B starting at A?

   (a) $14\frac{3}{8}$ **(b)** $18\frac{3}{8}$ **(c)** $17\frac{1}{4}$ **(d)** $10\frac{1}{4}$ **(e)** not given

21. By how much does $\frac{3}{8}$ of 12 exceed $\frac{1}{7}$ of 5?

   (a) $3\frac{11}{14}$ **(b)** $\frac{9}{2}$ **(c)** $\frac{5}{7}$ **(d)** $\frac{69}{14}$ **(e)** not given

22. If $\frac{3}{7}$ of Jim's weekly budget is $330. What is his weekly budget?

   (a) $110   **(b)** $330   **(c)** $990   **(d)** $770   **(e)** $440

H. Comparing Fractions, Complex Fractions, and Order of Operation

To find which of two fractions is the larger or smaller:

    a) if the two fractions being compared have the same denominator then compare numerators. The fraction with the smaller numerator is the smaller fraction.

    b) If the two fractions being compared have different denominators, convert both fractions to equivalent fractions with a common denominator; then compare numerators.

EXAMPLE 1:   In the following, determine the larger of the two fractions.

    (a) $\dfrac{5}{12}, \dfrac{7}{12}$ **(b)** $\dfrac{5}{12}, \dfrac{3}{8}$ **(c)** $\dfrac{13}{36}, \dfrac{19}{48}$ **(d)** $\dfrac{13}{18}, \dfrac{7}{12}$

SOLUTION:

    (a) since $\dfrac{5}{12}$ and $\dfrac{7}{12}$ have the same denominator and $7 > 5$, $\dfrac{7}{12} > \dfrac{5}{12}$

    (b) $\dfrac{5}{12}, \dfrac{3}{8}$ have different denominators, write both with common denominators.

        LCD of 12 and 18 = 24.

        $\dfrac{5 \cdot 2}{12 \cdot 2} = \dfrac{10}{24}, \quad \dfrac{3 \cdot 3}{8 \cdot 3} = \dfrac{9}{24}$ since $10 > 9$, then $\dfrac{10}{24} > \dfrac{9}{24}$

        So the larger fraction is $\dfrac{5}{12}$.

    (c) $\dfrac{13}{36}, \dfrac{19}{48}$ LCD of 36 and 48 = 144.

        $\dfrac{13 \cdot 4}{36 \cdot 4} = \dfrac{52}{144}, \quad \dfrac{19 \cdot 3}{48 \cdot 3} = \dfrac{57}{144}$ thus $\dfrac{57}{144} > \dfrac{52}{144}$

        So the larger fraction is $\dfrac{19}{48}$.

    (d) $\dfrac{13}{18}, \dfrac{7}{12}$ LCD of 18 and 12 = 36

        $\dfrac{13 \cdot 2}{18 \cdot 2} = \dfrac{26}{36}, \quad \dfrac{7 \cdot 3}{12 \cdot 3} = \dfrac{21}{36}$ thus $\dfrac{26}{36} > \dfrac{21}{36}$

        So the larger fraction is $\dfrac{13}{18}$.

A fraction that has at least one fraction in the numerator or in the denominator is called a **complex fraction**.

To simplify a complex fraction, we simplify the numerator and the denominator and then divide the simplified denominator into the simplified numerator.

EXAMPLE 2: Simplify

(a) $\dfrac{\frac{1}{4}+\frac{1}{3}}{\frac{1}{4}-\frac{1}{5}}$ (b) $\dfrac{1+\frac{4}{5}}{\frac{5}{8}}$

SOLUTION:

(a) $\dfrac{\frac{1}{4}+\frac{1}{3}}{\frac{1}{4}-\frac{1}{5}} = \dfrac{\frac{1\cdot3}{4\cdot3}+\frac{1\cdot4}{3\cdot4}}{\frac{1\cdot5}{4\cdot5}-\frac{1\cdot4}{5\cdot4}} = \dfrac{\frac{3}{12}+\frac{4}{12}}{\frac{5}{20}-\frac{4}{20}} = \dfrac{\frac{7}{12}}{\frac{1}{20}}$ add top fractions
   subtract bottom fractions

Now, divide the top fraction by the bottom fraction.

$$\frac{7}{12} \div \frac{1}{20} = \frac{7}{12} \cdot \frac{20}{1} = \frac{7}{3\cdot\cancel{4}} \cdot \frac{\overset{1}{\cancel{4}}\cdot5}{1} = \frac{35}{3}$$

(b) $\dfrac{1+\frac{4}{5}}{\frac{5}{8}} = \dfrac{\frac{1\cdot5}{1\cdot5}+\frac{4}{5}}{\frac{5}{8}} = \dfrac{\frac{5}{5}+\frac{4}{5}}{\frac{5}{8}} = \dfrac{\frac{9}{5}}{\frac{5}{8}}$ add top fractions

Divide the top fraction by the bottom fraction.

$$\frac{9}{5} \div \frac{5}{8} = \frac{9}{5} \cdot \frac{8}{5} = \frac{72}{25}$$

When evaluating or simplifying an expression involving more than one operation $(+, -, \times, \div)$, the order in which these operations are done can effect the final value.

EXAMPLE 3: Evaluate $\dfrac{5}{8} - \dfrac{1}{2} \cdot \dfrac{2}{3}$.

SOLUTION: If we subtract first we get

$$\frac{5}{8} - \frac{1}{2} = \frac{5}{8} - \frac{4}{8} = \frac{1}{8}$$

$$\frac{1}{8} \cdot \frac{2}{3} = \frac{1}{\cancel{2}\cdot4} \cdot \frac{\overset{1}{\cancel{2}}}{3} = \boxed{\frac{1}{12}}$$

However, if we multiply first, we get

$$\frac{5}{8} - \frac{1}{\cancel{2}} \cdot \frac{\overset{1}{\cancel{2}}}{3} = \frac{5}{8} - \frac{1}{3} = \frac{5\cdot3}{8\cdot3} - \frac{1\cdot8}{3\cdot8} = \frac{15}{24} - \frac{8}{24} = \boxed{\frac{7}{24}}$$

Which is correct?

To avoid such differences, when we have more than one operation, we perform operations in the following order:

1. Perform operations inside parentheses.
2. Do multiplication and division from left to right.
3. Do addition and subtraction last, again from left to right.

So according to these rules, $\dfrac{7}{24}$ is the correct answer to Example 3.

EXAMPLE 4: Simplify

(a) $\dfrac{3}{5} - \dfrac{1}{3} \div \dfrac{5}{6}$ (b) $\dfrac{3}{5} \cdot \dfrac{1}{2} + \dfrac{1}{5} \div \dfrac{2}{3}$ (c) $\dfrac{3}{5} \div \dfrac{6}{15} \cdot \dfrac{1}{2}$ (d) $\left(\dfrac{3}{5} - \dfrac{1}{3}\right) \div \dfrac{5}{6}$

SOLUTION:

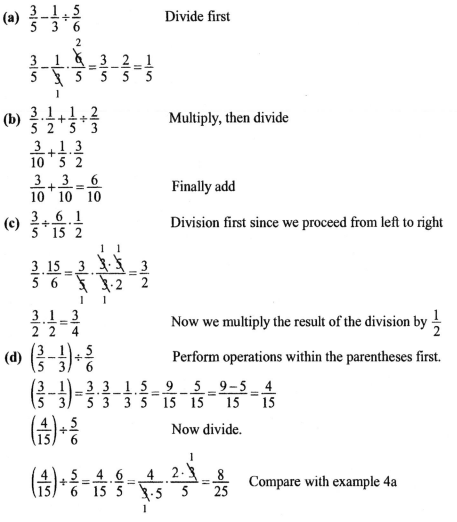

(a) $\dfrac{3}{5} - \dfrac{1}{3} \div \dfrac{5}{6}$ Divide first

$\dfrac{3}{5} - \dfrac{1}{3} \cdot \dfrac{\overset{2}{\cancel{6}}}{\underset{1}{\cancel{3}}} \cdot \dfrac{1}{5} = \dfrac{3}{5} - \dfrac{2}{5} = \dfrac{1}{5}$

(b) $\dfrac{3}{5} \cdot \dfrac{1}{2} + \dfrac{1}{5} \div \dfrac{2}{3}$ Multiply, then divide

$\dfrac{3}{10} + \dfrac{1}{5} \cdot \dfrac{3}{2}$

$\dfrac{3}{10} + \dfrac{3}{10} = \dfrac{6}{10}$ Finally add

(c) $\dfrac{3}{5} \div \dfrac{6}{15} \cdot \dfrac{1}{2}$ Division first since we proceed from left to right

$\dfrac{3}{5} \cdot \dfrac{15}{6} = \dfrac{3}{5} \cdot \dfrac{\overset{1}{\cancel{3}} \cdot \overset{1}{\cancel{5}}}{\underset{1}{\cancel{3}} \cdot 2} = \dfrac{3}{2}$

$\dfrac{3}{2} \cdot \dfrac{1}{2} = \dfrac{3}{4}$ Now we multiply the result of the division by $\dfrac{1}{2}$

(d) $\left(\dfrac{3}{5} - \dfrac{1}{3}\right) \div \dfrac{5}{6}$ Perform operations within the parentheses first.

$\left(\dfrac{3}{5} - \dfrac{1}{3}\right) = \dfrac{3}{5} \cdot \dfrac{3}{3} - \dfrac{1}{3} \cdot \dfrac{5}{5} = \dfrac{9}{15} - \dfrac{5}{15} = \dfrac{9-5}{15} = \dfrac{4}{15}$

$\left(\dfrac{4}{15}\right) \div \dfrac{5}{6}$ Now divide.

$\left(\dfrac{4}{15}\right) \div \dfrac{5}{6} = \dfrac{4}{15} \cdot \dfrac{6}{5} = \dfrac{4}{\underset{1}{\cancel{3}} \cdot 5} \cdot \dfrac{2 \cdot \overset{1}{\cancel{3}}}{5} = \dfrac{8}{25}$ Compare with example 4a

89

Exercises: Comparing Fractions, Complex Fractions, and Order of Operation

For 1 – 9, determine the larger fraction

1. $\frac{11}{40}, \frac{19}{40}$

2. $\frac{13}{14}, \frac{19}{21}$

3. $\frac{13}{16}, \frac{5}{8}$

4. $\frac{7}{24}, \frac{9}{33}$

5. $\frac{5}{12}, \frac{7}{15}$

6. $\frac{7}{9}, \frac{13}{15}$

7. $\frac{11}{16}, \frac{19}{24}$

8. $\frac{5}{12}, \frac{7}{18}$

9. $\frac{9}{14}, \frac{13}{21}$

For 10 – 14, simplify.

10. $\dfrac{\frac{1}{2}+\frac{3}{4}}{\frac{9}{10}}$

(a) $\frac{25}{18}$ (b) $\frac{5}{3}$ (c) $\frac{5}{9}$ (d) $\frac{5}{4}$ (e) not given

11. $\dfrac{\frac{3}{7}+\frac{1}{14}}{\frac{5}{6}}$

(a) $\frac{5}{3}$ (b) $\frac{3}{5}$ (c) $\frac{5}{9}$ (d) $\frac{5}{4}$ (e) not given

12. $\dfrac{\frac{2}{9}+\frac{4}{3}}{\frac{3}{2}+\frac{1}{3}}$

(a) $\frac{28}{33}$ (b) $\frac{14}{33}$ (c) $\frac{5}{12}$ (d) $\frac{12}{5}$ (e) not given

13. $\dfrac{\frac{5}{6}+\frac{1}{3}}{\frac{6}{8}+1}$

(a) $\frac{3}{2}$ (b) $\frac{4}{3}$ (c) $\frac{2}{3}$ (d) $\frac{3}{5}$ (e) not given

14. $\dfrac{\dfrac{3}{7}-\dfrac{5}{21}}{\dfrac{4}{9}-\dfrac{2}{18}}$

   (a) $\dfrac{3}{2}$ (b) $\dfrac{4}{3}$ (c) $\dfrac{2}{3}$ (d) $\dfrac{4}{7}$ (e) not given

For 15 – 19, simplify the using the order of operations rules.

15. $\dfrac{4}{5}+\dfrac{2}{7}\cdot\dfrac{21}{14}$

   (a) $\dfrac{43}{35}$ (b) $\dfrac{35}{43}$ (c) $\dfrac{84}{75}$ (d) $\dfrac{75}{84}$ (e) not given

16. $\dfrac{5}{6}\cdot\left(\dfrac{4}{3}-\dfrac{1}{6}\right)+\dfrac{7}{18}$

   (a) $\dfrac{7}{6}$ (b) $\dfrac{49}{36}$ (c) $\dfrac{6}{7}$ (d) $\dfrac{17}{18}$ (e) $\dfrac{1}{18}$

17. $\left(\dfrac{1}{2}+\dfrac{5}{4}\right)\div\dfrac{5}{8}$

   (a) $\dfrac{5}{14}$ (b) $\dfrac{5}{8}$ (c) $\dfrac{14}{5}$ (d) $\dfrac{3}{8}$ (e) $\dfrac{3}{64}$

18. $\dfrac{1}{2}+\left(\dfrac{1}{3}\div\dfrac{3}{4}\right)-\left(\dfrac{4}{5}\times\dfrac{5}{6}\right)$

   (a) $\dfrac{1}{3}$ (b) $\dfrac{7}{18}$ (c) $\dfrac{3}{18}$ (d) $\dfrac{5}{18}$ (e) not given

19. $\dfrac{2}{5}\div\dfrac{3}{8}\cdot\dfrac{6}{5}$

   (a) $\dfrac{16}{15}$ (b) $\dfrac{6}{5}$ (c) $\dfrac{75}{32}$ (d) $\dfrac{5}{2}$ (e) not given

20. If $\left(\dfrac{3}{4}-\dfrac{1}{3}\right)+\left(\dfrac{1}{2}+\dfrac{2}{3}\right)$ is calculated and the answer simplified, what is the denominator of the resulting fraction?

   (a) 24   (b) 12   (c) 6   (d) 4   (e) 3

Chapter 3

Decimals

SECTION A. Understanding Decimals and Decimal Fractions

A **decimal fraction** is a fraction whose denominator is a power of 10: 10, 100, 1000, 10,000, and so on. For example,

$$\frac{7}{10}, \frac{33}{100}, \text{ and } \frac{123}{1000} \text{ are examples of decimal fractions.}$$

The fraction $\frac{7}{10}$ is equal to .7 (read "point" seven) in **decimal notation**. Sometimes we write 0.7 (read zero "point" seven).

Note: 0.7 has the same value as .7

We can write decimal fractions in decimal notation. A fraction written in decimal notation is a **decimal form** or a **decimal number.**

Decimal Fraction	Decimal Form
$\frac{3}{10}$	0.3
$\frac{3}{100}$	0.03
$\frac{123}{1000}$	0.123
$\frac{12}{1000}$	0.012
$\frac{1234}{1000}$	1.234
$5\frac{67}{1000}$	5.067

In the above table, notice that the number of zeros in the denominator is equal to the number of decimal places to the right of the decimal point. For example,

$$\underbrace{\frac{12}{1000}}_{3 \text{ zeros}} = 0.\underbrace{012}_{3 \text{ decimal places}}$$

A mixed number, such as $5\frac{67}{1000} = 5.067$ has a **whole number part** to the left of the decimal point, and a **decimal part** to the right.

$$\underbrace{5}_{\text{whole number part}} . \underbrace{067}_{\text{decimal part}}$$

92

EXAMPLE 1: Write the decimal fractions in decimal notation.

    a. $\dfrac{34}{100}$ **b.** $\dfrac{34}{1000}$ **c.** $\dfrac{123}{10}$ **d.** $\dfrac{7}{1000}$ **e.** $\dfrac{4567}{100}$

SOLUTION:

    a. 0.34   **b.** 0.034   **c.** 12.3   **d.** 0.007   **e.** 45.67

Note: In part **b** we added a zero before the digits 34, since there are three zeros in the denominator, so we need three decimal places in the decimal form.

Note: In parts **c** (and in part **d**) the numerator is larger than the denominator so the fraction is larger than one. Notice that the decimal form, 12.3, has a whole number part.

EXAMPLE 2: Write the decimal form as a decimal fraction.

    a. 0.56   **b.** 65.975   **c.** 0.076   **d.** 5.0003

SOLUTION:

    a. $\dfrac{56}{100}$ **b.** $65\dfrac{975}{1000}$ **c.** $\dfrac{76}{1000}$ **d.** $5\dfrac{3}{10,000}$

The **word-name** for a decimal form is the same as the word-name for the corresponding decimal fraction.

The following chart lists the word names of some decimal fractions whose numerators are equal to one.

$\dfrac{1}{10} = 0.1$	$\dfrac{1}{100} = 0.01$	$\dfrac{1}{1000} = 0.001$	$\dfrac{1}{10,000} = 0.00001$
one tenth	one hundredth	one thousandth	one ten thousandth

The word name for $\dfrac{123}{1000} = 0.123$ is one hundred twenty-three thousandths.

EXAMPLE 3: Write the word names for the decimal fractions (and corresponding decimals).

    a. $\dfrac{34}{1000} = 0.034$ **b.** $\dfrac{3}{100} = 0.03$ **c.** $5\dfrac{67}{1000}$

SOLUTION:

    a. Thirty-four thousandths
    b. Three hundredths
    c. Five and sixty-seven thousandths

The following chart lists **place-values** to the right and the left of the decimal point.

Hundreds	Tens	Ones	Decimal Point	Tenths	Hundredths	Thousandths	Ten Thousandths
100	10	1	.	$\frac{1}{10} = 0.1$	$\frac{1}{100} = 0.01$	$\frac{1}{1000} = 0.001$	$\frac{1}{10,000} = 0.0001$

From the place-value chart we see that the decimal form 57.123 can be expanded as,

$$57.123 = 5(10) + 7(1) + 1\left(\frac{1}{10}\right) + 2\left(\frac{1}{100}\right) + 3\left(\frac{1}{1000}\right)$$
$$= 5(10) + 7(1) + 1(0.1) + 2(0.01) + 3(0.001).$$

EXAMPLE 4: Write the following decimals in **expanded form**.
    a. 1.245   **b.** 0.098

SOLUTION:

    a. $1.245 = 1 + 2\left(\frac{1}{10}\right) + 4\left(\frac{1}{100}\right) + 5\left(\frac{1}{1000}\right)$
                $= 1 + 2(0.1) + 4(0.01) + 5(0.001)$

    b. $0.098 = 9\left(\frac{1}{100}\right) + 8\left(\frac{1}{1000}\right)$
                $= 9(0.01) + 8(0.001)$

EXAMPLE 5: What does the digit 4 represent within the following decimal numbers?
    a. 4.123   **b.** 1.423   **c.** 0.04   **d.** 0.054

SOLUTION:

    a. 4 ones   **b.** $\frac{4}{10}$ **c.** $\frac{4}{100}$ **d.** $\frac{4}{1000}$

Exercises: Understanding Decimals and Decimal Fractions

For 1 – 8, write the decimal fractions in decimal notation.

1. $\dfrac{67}{100}$

2. $\dfrac{75}{1000}$

3. $\dfrac{456}{10}$

4. $68\dfrac{9}{100}$

5. $\dfrac{9845}{10}$

6. $\dfrac{35}{10,000}$

7. $\dfrac{567}{1000}$

8. $3\dfrac{7}{10,000}$

For 9 – 16, write the decimal forms as decimal fractions.

9. 0.23

10. 45.678

11. 0.0005

12. 0.987

13. 8.09

14. 4.0007

15. 0.067

16. 3.06

For 17 – 24, write the word name.

17. $\dfrac{9}{10}$

18. $\dfrac{78}{100}$

19. $\dfrac{5}{1000}$

20. $34\dfrac{45}{100}$

21. 0.8

22. 0.56

23. 0.0003

24. 57.087

For 25 – 32, write in expanded form.

25. 5.36 **26.** 368.75 **27.** 0.56

28. 0.03 **29.** 0.078 **30.** 0.00056

31. 0.2345 **32.** 0.567

For 33 – 36, what does the 5 represent in the following decimal numbers?

33. 0.567

34. 0.956

35. 46.935

36. 5.89

37. In the decimal number 0.33, the digit 3 on the right is _____ times the digit 3 on the left.

38. In the decimal number 0.33, the digit 3 on the left is _____ times the digit 3 on the right.

39. In the decimal number 0.3003, the digit 3 on the right is _____ times the digit 3 on the left.

40. In the decimal number 0.3003, the digit 3 on the left is _____ times the digit 3 on the right.

SECTION B. Ordering and Comparing Decimals

Which is larger 0.035 or 0.047?
We can compare decimals that have the same number of digits to the right of the decimal point,
by comparing the numbers without regard to their decimal points.
For example, 0.035 < 0.047 since 35 < 47.
If the decimals do not have the same number of digits to the right of the decimal point, then fill in
with zeros on the right.
For example, if we compare 0.1 with 0.013, add two zeros to the right of 0.1. Now compare
0.100 with 0.013.

EXAMPLE 1:   Replace the ? with the symbol: < or >.
                        0.01 ? 0.0099

SOLUTION:
First add two zeros on the right of the first decimal number.
0.0100 ? 0.0099
0.0<u>100</u> > 0.00<u>99</u> since 100 > 99.

Note: Adding two zeros onto the end of a decimal number is the same as multiplying the top and
bottom of the corresponding decimal fraction by 100. Thus, .01 = .0100 since

$$\frac{1}{100} = \frac{100}{10,000}$$

Note: Removing trailing zeros on the right of the decimal point is the same as dividing the top
and bottom of the corresponding decimal fraction by a power of 10. For example 6.9500 is the
same as 6.95.

EXAMPLE 2:   Arrange in order from smallest to largest.
                        0.05, 0.032, 0.1, 0.123, 0.0099

SOLUTION: First add right-hand zeros so that all the numbers have the same number of places
to the right of the decimal point.

    0.0500   500
    0.0320   320
    0.1000   1000   Compare these values.
    0.1230   1230
    0.0099   99

The numbers in order are: 0.0099, 0.032, 0.05, 0.1, 0.123.

EXAMPLE 3: Place the following decimal numbers on the number line.
0.30, 0.65, 0.15, 0.7

SOLUTION:

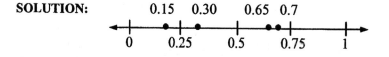

C. Rounding Decimals

We can approximate a decimal number with a "rounded" number. For example, 3.423 is approximately equal to 3.4 and 5.29 is approximately equal to 5.3. We say that the numbers 3.423 and 5.3 have been "rounded to the nearest tenth". In this case the tenth digit is the called the **round-off digit**.

Procedure to Round Decimals
1. If the digit to the right of the round-off digit is less than 5, then drop the digits to the right of the round-off digit.
2. If the digit to the right of the round-off digit is greater than or equal to 5, then increase the round off digit by one, and drop the digits to the right of the round-off digit.

EXAMPLE 1: Round to the nearest hundredth.
      a. 15.3429   **b.** 0.13822   **c.** 7.2352   **d.** 0.198

SOLUTION:
    a. The round-off digit is the hundredth digit, or the 4. The digit to the right is 2. Since 2 is less than 5 we do not change the round-off digit. Therefore 15.3429 rounded to the nearest hundredth is 15.34.
    b. The round-off digit is 3. The digit to the right is greater than 5. Therefore 0.13822 rounded to the nearest hundredth is 0.14.
    c. The round-off digit is 4. The number to the right is 5. Therefore 7.2352 rounded to the nearest hundredth is 7.24.
    d. The round-off digit is 9. The digit to the right is 8. Therefore 0.198 rounded to the nearest hundredth is 0.20.

Exercises: Ordering and Comparing Decimals and Rounding

For 1 – 6, replace the ? by a < or > symbol.

1. 0.57 ? 0.571   **2.** 0.050 ? 0.15   **3.** 0.015 ? 0.0143

4. 1.099 ? 1.9   **5.** 0.099 ? 0.1   **6.** 0.0123 ? 0.12

For 7 – 8, write the decimal numbers in order of smallest to largest.

7. 0.099, 0.110, 0.00999, 0.109, 0.19

8. 3.123, 3.0123, 30.1, 3.099, 3.90

9. If the zero in 4.10 were omitted, would the value of the number change? Explain.

10. If the zero in 0.5 were omitted, would the value of the number change?

11. If a zero were added to the right of 81.9, would the value of the number change?

12. If a zero is placed between the decimal point and the 5 on .5, does the value of the number change? Explain.

For 13 – 16, what are the missing numbers?

13. 0.08, 0.09, _____, 0.11, _____, _____

14. 2.06, 2.08, _____, _____

15. 1.02, 1.01, 1, _____, _____

16. 0.5, 1, 1.5, _____, _____, _____, _____

100

17. Which numbers are equal to 0.2?   0.02, .2, .20, 0.20, $\frac{2}{10}$, $\frac{20}{100}$, 2.0, two tenths, 20 hundredths, 200 thousandths

18. How man 0.1's are there in 1.0?

19. How many 0.5's are there in 2.0?

For 20 – 27, round to the nearest given place.

   20. 3.234, tenth

   21. 3.234, hundredth

   22. 5.3426, thousandth

   23. 6.75, tenth

   24. 3.006, hundredth

   25. 4.099, hundredth

   26. 2.999, hundredth

   27. 3.52, whole number

28. Round to the nearest whole number, and then add.   0.98 + 2.05 + 0.09 + 1.5

29. Kate went shopping for groceries. As the cashier rang up her items, Kate approximated the total by rounding the price of each item to the nearest dollar, and adding. What was Kate's approximate total for the following prices: $1.05, $2.50, $.98, $1.39, $2.75, $3.49?
   (a) $12   **(b)** $13   **(c)** $14   **(d)** $11   **(e)** $10

SECTION D. Adding and Subtracting Decimal Numbers

We add decimal numbers with two places the same way we add money. For example,

$3.15   Line up the decimal places, and add.
$2.50

$5.65

Procedure to Add or Subtract Decimals
1. Write the numbers vertically and line up the decimal points. You may add extra zeros so that all the numbers have the same number of decimal places to the right of the decimal point.
2. Add (or subtract) all the digits within the same column.
3. Place the decimal point of the sum (or difference) in line with the decimal points of all the numbers added (or subtracted).

EXAMPLE 1: Add $0.33 + 5.27 + 2.45$

SOLUTION:

    1 1   the carried ones
    0.33
    5.27
    2.45

    8.05

EXAMPLE 2: $5.1 - 0.02$

SOLUTION:

     5.10   Fill in one zero on the right.
    -0.02

     5.08

Exercises: Adding and Subtracting Decimal Numbers

For 1 – 4, add the following decimal numbers.

1. $0.020 + 23.5 + 1.003$

2. $1.012 + 35.6 + 4.702$

3. $12.8 + 0.71 + 0.091$

4. $505.5 + 0.0005 + 50.005$

For 5 – 13, subtract the following decimal numbers.

5. $5.5 - 0.2$ **6.** $5.1 - 0.3$ **7.** $1.0 - 0.1$

8. $5.02 - 0.3$ **9.** $0.99 - 0.05$ **10.** $1 - 0.5$

11. $1 - 0.01$ **12.** $1 - 0.001$ **13.** $8 - 0.02$

For 14 – 16, add the following decimal numbers.

14. $\begin{array}{r} 3.45 \\ +18.036 \\ \hline \end{array}$ **15.** $\begin{array}{r} 28.075 \\ +\ 6.92 \\ \hline \end{array}$ **16.** $\begin{array}{r} 352.976 \\ +\ 48.95 \\ \hline \end{array}$

For 17 – 19, subtract the following decimal numbers.

17. $\begin{array}{r} 97.034 \\ -15.372 \\ \hline \end{array}$ **18.** $\begin{array}{r} 21.575 \\ -\ 3.02 \\ \hline \end{array}$ **19.** $\begin{array}{r} 89.27 \\ -\ 4.003 \\ \hline \end{array}$

20. Subtract 0.03 from 10.

21. 55 is how much more than 27.3?

22. What must you add to 43.78 to get 50?

23. How much more than $5\frac{3}{4}$ is 5.87?

24. Nick's yard is 43.25 feet by 38.3 feet, by 28.2 feet by 25.75 feet. What is the perimeter of his yard?
  (a) 556.24   **(b)** 109.8   **(c)** 119.5   **(d)** 135.5   **(e)** 4078.45

25. Gary bought 3.76 pounds of veal, 11.7 pounds of beef and $21\frac{4}{5}$ pounds of chops for a school barbecue. How many pounds of meat did he buy?
  (a) 35.8   **(b)** 65.6   **(c)** 37.26   **(d)** 56.7

26. In the morning the odometer on John's car registered 5287.5 miles. At night it showed 5363.2 miles. How far did John drive during the day?
  (a) 76.4   **(b)** 67.9   **(c)** 46.9   **(d)** 85.5   **(e)** 75.7

27. A baby girl's birth weight was 3.61 kilograms. At the baby's three month checkup she weighed 4.57 kilograms. How much weight did she gain?
  (a) 1.01 kg   **(b)** 1.26 kg   **(c)** 0.58 kg   **(d)** 0.96 kg   **(e)** 1.24 kg

28. Four runners ran 100 yards in 13.5 seconds, 18.2 seconds, 15.8 seconds, and 13.9 seconds. What was their total running time?
  (a) 61.4   **(b)** 78.4   **(c)** 89.2   **(d)** 56.3   **(e)** 87.7

29. Two runners ran the 100 yard dash in 12.8 seconds and 11.07 seconds. The first runner's time was how much more than the second runner's time?
  (a) 2.07   **(b)** 1.56   **(c)** 2.01   **(d)** 2.31   **(e)** 1.73

SECTION E. Multiplication and Division of Decimal Numbers

Multiplication of decimal numbers
Let us multiply 0.25×0.55 by multiplying the equivalent decimal fraction,

$$0.25 \times 0.55 = \frac{25}{100} \times \frac{55}{100} = \frac{1375}{10,000} = 0.1375$$

↑

> The number of decimal places in the product
> is equal to the total number of decimal places
> in the numbers being multiplied.

The above problem is an example of the procedure for multiplying decimal numbers.

The procedure for Multiplying Decimal Numbers
1. Multiply the two numbers without regard to their decimal points.
2. The product contains the number of decimal places equal to the total number of decimal places in the numbers being multiplied.

EXAMPLE 1: Multiply 0.345×0.0032.

SOLUTION:
First multiply the two numbers without regard to the decimal points.

$$
\begin{array}{r}
0.345 \\
\times\, 0.0032 \\
\hline
690 \\
1035 \\
\hline
11040
\end{array}
$$

0.345 has 3 decimal places to the right of zero
0.0032 has 4 decimal places to the right of zero
The product has 3 + 4 = 7 decimal places to the right of zero.
The product is 0.0011040.

Multiplication by powers of 10
We saw in the chapter on Whole Numbers, that multiplication of a whole number by 10 will add a zero onto the end of the number. If we multiply a decimal number by 10, then the decimal place moves one place to the right. For example,

Obtain the product 3.75×10 :
    First: multiply $375 \times 10 = 3750$
    Second: position the decimal point so that the product has 2 decimal places.
So, $3.75 \times 10 = 3\,7.50$.

Multiplication of a decimal number by a power of 10, moves the decimal point over to the right by as many places as the number of zeros in the power of 10.

EXAMPLE 2: Multiply by the power of 10.
    a. 3.45×10 **b.** 0.098×100 **c.** 45.3567×1000

SOLUTION:
    a. 34.5   **b.** 9.8   **c.** 45,356.7

EXAMPLE 3: Multiplication by one-tenth, one-hundredth, and one-thousandth
    a. 56.75×0.1 **b.** 0.78×0.01 **c.** 4.236×0.001

SOLUTION:
    a. 5.675   **b.** 0.0078   **c.** 0.004236

From the above example we see that multiplication by one-tenth moves the decimal point one place to the left, multiplication by one-hundredth moves the decimal point two places to the left, and multiplication by one-thousandth moves the decimal point three places to the left.

Division of a decimal number by a whole number
Recall that when we divide whole numbers, we continue the "long division" procedure until there are no numbers left to "bring down" and we are left with a remainder (the remainder could be zero). When we divide decimal numbers we no longer present the quotient together with a remainder. Rather, we obtain a quotient with more precision by continuing the division until we get the number of decimal places that we want.

To divide a decimal number by a whole number place the decimal point in the quotient, above the decimal point in the dividend. Then proceed as if the dividend were a whole number.

106

EXAMPLE 4: Divide $27\overline{)93.15}$.

SOLUTION:

$$
\begin{array}{r}
3.45 \\
27\overline{)93.15} \\
\underline{81} \\
121 \\
\underline{108} \\
135 \\
\underline{135} \\
0
\end{array}
$$

If we do not have a remainder of zero, then we can add zeros to the right of the dividend. Sometimes we get a remainder of zero and stop the long division. If we do not get a remainder of zero, then we continue the division until the desired number of decimal places is in the quotient.

EXAMPLE 5: Divide and round the quotient to the nearest thousandth

$$28\overline{)430.3}$$

SOLUTION:

$$
\begin{array}{r}
15.3678 \\
28\overline{)430.3000} \\
\underline{28} \\
150 \\
\underline{140} \\
103 \\
\underline{84} \\
190 \\
\underline{168} \\
220 \\
\underline{196} \\
240 \\
\underline{224} \\
16
\end{array}
$$

We need 4 decimal places to round to the nearest thousandth.

Rounded: 15.368

107

Division of a decimal number by a decimal number

In the chapter on Whole Numbers, we saw that there are many ways to express a division problem. Let us rewrite the division problem, $0.12 \div 0.03$ as the "fraction" $\dfrac{0.12}{0.03}$.

By multiplying numerator and denominator by 100 we form an equivalent fraction without decimal points in the numerator or denominator.

$$\frac{0.12}{0.03} = \frac{(0.12)(100)}{(0.03)(100)} = \frac{12}{3} = 4.0$$

Now, let us rewrite the same example as a "long division" problem. We have seen that multiplication of numerator and denominator by 100 is equivalent to moving the decimal points in dividend and divisor two places to the right.

$$0.03\overline{)0.12} = 3\overline{)12.0}^{\,4.0}$$ Move the decimal point of the divisor and dividend two places to the right.

Procedure for division of decimal numbers

1. Eliminate the decimal point in the divisor by moving it to the right of the right most digit.
2. Move the decimal point in the dividend to the right, the same number of places that the decimal point in the divisor was moved. You may add zeros onto the end of the dividend, to the right of the decimal point.
3. Position the decimal point in the quotient above the decimal point in the divisor.

EXAMPLE 6: Divide $1.568 \div 0.28$

SOLUTION:

$$0.28\overline{)1.568}$$ Move the decimal points two places to the right.

$$
\begin{array}{r}
5.6 \\
28\overline{)156.8} \\
\underline{140} \\
168 \\
\underline{168} \\
0
\end{array}
$$

Check: $(0.28)(5.6) = 1.568$

108

Division by powers of 10

We saw that when we multiply a decimal number by a power of 10 the decimal shifts to the right by as many places as there are zeros in the power of 10. For example,

$$(3.456)(100) = 345.6$$

When we divide a decimal number by a power of 10, the decimal point is shifted to the left by as many places as there are zeros in the power of 10. For example,

$$\frac{345.6}{100} = 3.456$$

EXAMPLE 7: Divide by the power of 10.
      a. $453.234 \div 100$ **b.** $0.037 \div 1000$

SOLUTION:
      a. $453.234 \div 100 = 4.53234$
      b. $0.037 \div 1000 = 0.000037$

Exercises: Multiplication and Division of Decimal Numbers

1. A number with 3 decimal places is multiplied by a number with 5 decimal places. The product will have _____ decimal places.

For 2 – 10, multiply.

2. 0.08×0.003 3. 0.006×0.00002 4. 7×0.0005

5. 0.25×0.03 6. 4.5×2 7. 0.35×0.003

8. 0.54×0.32 9. 0.36×1.5 10. 0.000025×0.025

For 11 – 13, multiply.

11. 383.02
   $\times\ 56.5$

12. 38.26
   $\times\ 35.1$

13. 52.03
   $\times\ 5.7$

For 14 – 22, multiply by the given power of 10.

14. 12.345×10 15. 4.567×100 16. 34×100

17. 65.8×100 18. 3.2×1000 19. 0.00234×1000

20. 4.25×10^2 21. 3.07×10^4 22. 0.00007×10^5

For 23 – 28, divide.

23. $1.2 \div 3$ 24. $0.35 \div 7$ 25. $2.7 \div 3$

26. $12 \div 0.3$ 27. $35 \div 0.07$ 28. $15 \div 0.003$

For 29 – 36, divide. Continue the division until you get a remainder of zero.

29. $52\overline{)44.72}$ **30.** $7\overline{)3.661}$ **31.** $28\overline{)1.26}$

32. $16\overline{)1.36}$ **33.** $2.3\overline{)124.2}$ **34.** $0.35\overline{)164.5}$

35. $3.8\overline{)28.5}$ **36.** $0.25\overline{)8.1}$

For 37 – 40, divide by the given power of 10.

37. $28.123 \div 10$ **38.** $153.6 \div 100$ **39.** $356.2 \div 1000$

40. $7.3 \div 100$

For 41 – 43, divide and round to the nearest hundredth.

41. $4\overline{)263.82}$ **42.** $1.7\overline{)20.8}$ **43.** $29\overline{)4.073}$

For 44 – 46, divide and round to the nearest thousandth.

44. $8\overline{)0.2019}$ **45.** $0.69\overline{)8.45}$ **46.** $0.87\overline{)79.40}$

For 47 – 48, a freight company transports containers for a fee that depends on the number of miles the container travels.

47. Suppose the charge is \$2.68 per mile. What is the cost of sending a container 150 miles?
     (a) \$345   **(b)** \$554.68   **(c)** \$45.68   **(d)** \$402

48. If the company charges \$500 for a 235 mile trip, what is the cost per mile?
     (a) \$55.32   **(b)** \$2.13   **(c)** \$0.35   **(d)** \$2.35   **(e)** \$22

111

For 49 – 50, a warehouse contains stacks of sheet metal.

49. If a single sheet of metal is 0.27 centimeters thick, how many centimeters high would a stack of 137 sheets be?
   (a) 25.3   **(b)** 4.5   **(c)** 89   **(d)** 37   **(e)** 34.2

50. A stack of sheet metal is 40 centimeters high. If there are 160 sheets in the stack, how thick is each sheet?
   (a) 0.25   **(b)** 0.13   **(c)** 2.5   **(d)** 1.3

For 51 – 52, Stan went on a trip.

51. If Stan used 80 gallons of gasoline and averaged 23.26 miles to the gallon, then how far did he travel?
   (a) 823.26 mi **(b)** 1032.6 mi **(c)** 1860.8 mi **(d)** 186.08 mi **(e)** 103.26 mi

52. Suppose Stan traveled 265 miles and used 21.2 gallons of gasoline. On the average, how many miles did he average per gallon of gasoline?
   (a) 12.5   **(b)** 5565   **(c)** 26.5   **(d)** 42   **(e)** 21

53. A family of two adults and three children are going on a trip from New York to Washington. They can fly for $130 per person, or go by train for $89.50 per person. How much do they save if they take the train?
   (a) $125   **(b)** $202.50   **(c)** $447.50   **(d)** $650   **(e)** $250.50

For 54 – 55, Karen is baking cakes for a cake sale. She estimates that each cake costs her $2.50 in ingredients. She intends to sell her cakes for $5.25 each.

54. What will be her profit if she sells 24 cakes?
   (a) $25   **(b)** $2.75   **(c)** $60   **(d)** $126   **(e)** $66

55. How many cakes will she have to sell to make a profit of $110 ?
   (a) 25   **(b)** 40   **(c)** 60   **(d)** 66   **(e)** 50

For 56 – 58, Max buys gasoline either close to his house or a few blocks away.

56. At the gas station closest to Max's house, a gallon of gasoline is $1.69 a gallon. Max's tank holds 23 gallons. How much does it cost to fill his tank?
    (a) $46.02   **(b)** $44.34   **(c)** $48.04   **(d)** $13.09   **(e)** $38.87

57. If Max goes to the farther gas station he can fill his tank for $35.00. What is the price per gallon, to the nearest penny, at the farther gas station?
    (a) $1.37   **(b)** $1.55   **(c)** $1.43   **(d)** $1.52   **(e)** $1.25.

58. How much does Max save if he goes to the farther gas station and fills his empty tank?
    (a) $3.91   **(b)** $1.23   **(c)** $4.55   **(d)** $12.30   **(e)** $3.22

For 59 – 61, calculate the following quantities that compare the numbers 4.2 and 2.8.

59. The number 4.2 is _____ larger than the number 2.8?

60. 4.2 times 2.8 is _____.

61. The number 4.2 is _____ times the number 2.8?

62. The HQ Laser Jet super fast printer, produces 5.5 pages in 15 seconds. The HQ Laser medium fast printer, produces 5 pages in 30 seconds. The output of the super fast printer is how many times the output of the medium fast printer?
    (a) 2.2   **(b)** 1.1   **(c)** 0.5   **(d)** 3.1   **(e)** 4.5

63. Five friends shared the cost of renting a van. The cost was $23.50 per person. If seven friends shared the van, then how much would each person save?
    (a) $3.45   **(b)** $4.80   **(c)** $6.71   **(d)** $16.79   **(e)** $6.79

SECTION F. Converting Fractions to Decimals and Decimals to Fractions

Converting Decimals to Fractions

We convert a decimal number to a fraction by writing the decimal number in the form of a decimal fraction, and reducing if possible. For example,

$$0.25 = \frac{25}{100} = \frac{1}{4}$$

EXAMPLE 1:   Convert the decimal number to a fraction or mixed number.
      a. 78.127   **b.** 0.0087   **c.** 0.15

SOLUTION:

    a. $78.127 = 78\frac{127}{1000}$

    b. $0.0087 = \frac{87}{10,000}$

    c. $0.15 = \frac{15}{100} = \frac{3}{20}$ (reduced)

The table below has some frequently occurring decimal numbers and their equivalent fractions. You should be able to recognize these equivalencies.

Decimal Form	Fraction Form
0.5	$\frac{1}{2}$
0.25	$\frac{1}{4}$
0.2	$\frac{1}{5}$
0.4	$\frac{2}{5}$
0.75	$\frac{3}{4}$
0.125	$\frac{1}{8}$
1.5	$1\frac{1}{2} = \frac{3}{2}$

114

Converting Fractions to Decimals

Every fraction can be thought of as a division problem where the denominator divides the numerator. We convert a fraction into a decimal number by dividing the denominator into the numerator. For example,

$$\frac{1}{8} = 8\overline{)1.000}^{0.125}$$

We add zeros onto the end of the dividend to obtain a quotient with decimal places. Sometimes we continue the division until we get a remainder of zero. Sometimes we continue the division until we observe a repeating pattern in the quotient or until we obtain the number of decimal places that we want.

Every fraction has a decimal equivalent that either has finitely many decimal places, or has infinitely many decimal places that eventually form a repeating pattern.

EXAMPLE 2: Find the decimal equivalent of the fraction $\frac{7}{8}$.

SOLUTION:

$$
\begin{array}{r}
0.875 \\
8\overline{)7.000} \\
\underline{64} \\
60 \\
\underline{56} \\
40 \\
\underline{40} \\
0
\end{array}
$$

EXAMPLE 3: Find the decimal equivalent of $\frac{12}{99}$.

SOLUTION:

$$
\begin{array}{r}
0.1212 \\
99{\overline{\smash{\big)}\,12.0000}} \\
\underline{99} \\
210 \\
\underline{198} \\
120 \\
\underline{99} \\
210 \\
\underline{198} \\
12
\end{array}
$$

Notice that the decimal places form a repeating pattern.

$\frac{12}{99} = 0.121212...$

EXAMPLE 4: Find the decimal equivalent of $\frac{2}{7}$ rounded to the nearest thousandth.

SOLUTION:

$$
\begin{array}{r}
0.2857 \\
7{\overline{\smash{\big)}\,2.0000}} \\
\underline{14} \\
60 \\
\underline{56} \\
40 \\
\underline{35} \\
50 \\
\underline{49} \\
1
\end{array}
$$

We need 4 decimal places to round to the nearest thousandth.

Rounded: 0.286

To summarize, every fraction can be written as a decimal number. The decimal places either "terminate" (there are finitely many decimal places) or go on forever and after a while form a repeating pattern.

Every decimal number with finitely many decimal places can be written as a fraction. It is also possible to convert a decimal number having infinitely many "repeating" decimal places into a fraction, but that is beyond the scope of this book.

Exercises: Converting Fractions to Decimals and Decimals to Fractions

For 1 – 6, learn the table of decimal and fraction equivalencies found in this section. Answer exercises 1 – 6 based on your knowledge of the table.

1. $\frac{1}{4} =$ 　　　　　　 2. $\frac{3}{4} =$ 　　　　　　 3. $\frac{1}{5} =$

4. $0.2 =$ 　　　　　　 5. $0.4 =$ 　　　　　　 6. $1.5 =$

For 7 – 11, change the decimal forms to decimal fractions, and reduce the fractions.

7. 0.6 　　　　　　 8. 0.12 　　　　　　 9. 0.15

10. 0.375 　　　　　　 11. 2.25

For 12 – 27, write as an equivalent decimal. If a repeating decimal is obtained, use notation such as $0.\overline{12}$ or $0.12\overline{34}$.

12. $\frac{4}{5}$ 　　　　　　 13. $\frac{3}{8}$ 　　　　　　 14. $\frac{23}{50}$

15. $\frac{11}{20}$ 　　　　　　 16. $\frac{13}{25}$ 　　　　　　 17. $\frac{7}{40}$

18. $\frac{13}{50}$ 　　　　　　 19. $\frac{7}{5}$ 　　　　　　 20. $\frac{17}{99}$

21. $\frac{7}{11}$ 　　　　　　 22. $\frac{5}{18}$ 　　　　　　 23. $\frac{5}{6}$

24. $\frac{8}{27}$ 　　　　　　 25. $\frac{5}{33}$ 　　　　　　 26. $\frac{1}{27}$

27. $5\frac{7}{12}$

117

For 28 – 35, round to the nearest thousandth.

28. $\frac{2}{7}$ 29. $\frac{3}{14}$

30. $\frac{5}{13}$ 31. $\frac{15}{17}$

32. $\frac{5}{26}$ 33. $\frac{3}{23}$

34. $\frac{5}{21}$ 35. $\frac{3}{7}$

36. Add: $2.34 + 3\frac{2}{5} + 5\frac{1}{8}$

37. An electrician is drilling a hole for a cable that is 0.6 inch in diameter. He chooses a $\frac{5}{8}$ inch drill bit. Is the drill bit large enough? How much larger or smaller is it than the cable?

SECTION G. Solving Application Problems

Read each problem carefully and understand what quantity you are being asked to find. Gather the information given in the problem. Determine which operations are needed.

EXAMPLE 1: Benny sells baseball cards to his classmates. He buys the cards for $1.40 each and sells them for $1.75 each.
- **a.**   What is his profit per card?
- **b.**   If he sells 25 cards, then what is his profit?
- **c.**   Suppose Benny buys a box of 18 cards for $20.70, what was his cost per card?

SOLUTION:
- **a.**   Subtract 1.75 – 1.40 = $0.35.   He earns $0.35 per card.
- **b.**   Multiply 25 × 0.35 = $8.75.   Benny's profit earned selling 25 cards.
- **c.**   To obtain the cost per card we must divide $20.70 ÷ 18 = $1.15, the cost per card.

EXAMPLE 2:   A Laser printer produces 15 pages per minute. A typist types one page in four minutes.
- **a.**   What is the difference between the per minute output of the printer and the per minute output of the typist?
- **b.**   The output of the printer is how many times that of the typist?

SOLUTION:
- **a.**   Notice that the output of the printer is given for one minute and the output of the typist is given for four minutes. First we must find the one minute output of the typist.

   The typist produces $1 \div 4 = \frac{1}{4}$ page per minute.

   The output difference is $15 - \frac{1}{4} = \frac{15}{1} \cdot \frac{4}{4} - \frac{1}{4} = \frac{60}{4} - \frac{1}{4} = \frac{59}{4}$ pages per minute;

   expressed as a decimal, $59 \div 4 = 14.75$ pages per minute.
- **b.**   The output of the printer (15 pages per minute) is how many times the output of the typist ($\frac{1}{4}$ page per minute). That is, if $(15)(x) = \frac{1}{4}$, what is x? So, we must divide 15 by $\frac{1}{4}$, or by the equivalent decimal form 0.25.

$$0.25\overline{)15} \quad \text{Move the decimal points over two places ot the right}$$
   in both the divisor and the dividend.

$$\begin{array}{r} 60.0 \\ 25\overline{)1500.0} \end{array}$$

So, the output of the printer is 60 times that of the typist.

EXAMPLE 3: David works in a restaurant. He earns $8.35 per hour. He earns $1\frac{1}{2}$ times that wage for any hours beyond 8 hours. Suppose he works 12 hours. What are his earnings?

SOLUTION:
For the first 8 hours his earnings are $(8)(8.35) = \$66.80$.

For the next $12 - 8 = 4$ hours his earnings are $(4)\left(1\frac{1}{2}\right)(8.35)$. Change $1\frac{1}{2}$ to the equivalent decimal form, 1.5, and multiply $(4)(1.5)(8.35) = \$50.10$. So he earns $\$66.80 + \$50.10 = \$116.90$ for 12 hours of work.

EXAMPLE 4: If one U.S. dollar is worth 1.20 Canadian dollars, and one Canadian dollar is worth 10.50 Mexican Peso's, then,
   a.   what is one U.S. dollar worth in Mexican Pesos?
   b.   what are 20 U.S. dollars worth in Mexican Pesos?

SOLUTION:
   a.   One U.S. dollar is 1.20 Canadian dollars.
        One Canadian dollar is worth 10.50 Mexican Pesos.
        So, 1.20 Canadian dollars is worth $(1.20)(10.50) = 12.60$ Mexican Pesos.
   b.   If one U.S. dollar is worth 12.60 Mexican Pesos, then 20 U.S. dollars are worth
        $(20)(12.60) = 252.00$ Mexican Pesos.

EXAMPLE 5: An amusement park has an entrance fee and an equal charge for each ride. On weekends the entrance fee is the same as during the week but the charge for each ride is more.
   a.   Gordon went to the park on a weekday and rode on only one ride. He paid $3.05 (including the entrance fee). On Sunday Gordon went to the park and again rode only one ride. He paid $3.30 (including the entrance fee). How much more is the charge for each ride on a weekend than on a weekday?
   b.   On Wednesday Sam and Gordon went to the park. Sam rode one ride and paid $3.05 (including entrance fee) and Gordon rode two rides and paid $3.80 (including entrance fee). What is the charge for each ride?
   c.   If Sam and Gordon went to a different park and Sam paid $6.30 for 3 rides (including entrance fee), and Gordon paid $8.70 for 5 rides (including entrance fee). What was the charge for each ride?

SOLUTION:
   a.   The difference $3.30 - $3.05 = $0.25 is the amount the charge for each ride increases on weekends.
   b.   The difference $3.80 - $3.05 = $0.75 is the amount charged for one ride.
   c.   The difference $8.70 - $6.30 = $2.40 is the amount charged for two rides. The amount charged for each ride is $2.40 \div 2 = \$1.20$.

EXAMPLE 6: A Fahrenheit temperature is equal to 1.8 times the equivalent Celsius temperature plus 32.

    a. Find the Fahrenheit temperature when the Celsius temperature is 20 degrees.

    b. Find the Celsius temp when the Fahrenheit temperature is 50 degrees.

SOLUTION:

    a. Multiply $(1.8)(20) = 36$. Now add $36 + 32 = 68$ degrees Fahrenheit.

    b. To find Celsius temperatures we do the opposite operations (subtraction is the opposite of addition and division is the opposite of multiplication) and in the opposite order.

        First subtract, $50 - 32 = 18$.

        Now divide, $18 \div 1.8 = 10$ degrees Celsius.

Exercises: Solving Application Problems

For 1 – 2, Sue is baking cakes for a cake sale. She estimates that each cake costs her $2.25 to make. She intends to sell her cakes for $4.00 each.

   1. If she sells 25 cakes, what will her profit be?
       (a) $56.25   **(b)** $100   **(c)** $50.00   **(d)** $43.75   **(e)** $156.25

   2. How many will she have to sell to make a profit of $21.00?
       (a) 10   **(b)** 7   **(c)** 9   **(d)** 18   **(e)** 12

For 3 – 4, a cracker is $\frac{1}{8}$ of an inch in thickness.

   3. If there are 22 crackers stacked one on top of the other. How high is the stack of crackers?
       (a) 176 in.   **(b)** 30 in.   **(c)** 2.75 in.   **(d)** 2 in.   **(e)** 3 in.

   4. Suppose the stack of crackers is 0.75 inches high, then how many crackers are in the stack?
       (a) 9   **(b)** 20   **(c)** 5   **(d)** 6   **(e)** 10

For 5 – 6, a laser printer prints 125 characters in one second. A typist types 2.5 characters in one second.

   5. What is the difference between the one second output of the printer and the one second output of the typist?
       (a) 122.5   **(b)** 1.25   **(c)** 10   **(d)** 50   **(e)** 12

   6. The output of the printer is how many times the output of the typist?
       (a) 10   **(b)** 5   **(c)** 50   **(d)** 30   **(e)** 122.5

For 7 – 8, a laser printer prints 175 characters in one second. A typist types 150 characters in one minute.

7. What is the difference between the one second output of the printer and the one second output of the typist?
   (a) 100   (b) 25   (c) 50   (d) 172.5   (e) 125

8. The output of the printer is how many times the output of the typist?
   (a) 70   (b) 1.2   (c) 0.85   (d) 85   (e) 2

9. How much does 350 gallons of water cost at $0.025 a gallon?
   (a) $35   (b) $87.50   (c) $1400   (d) $75   (e) $8.75

10. If 500 gallons of water cost $17.50, how much does one gallon cost?
    (a) $0.25   (b) $0.035   (c) $0.35   (d) $0.29   (e) $2.86

11. A person drinks an average of 1.32 quarts of water each day. If a certain household of 5 people has a supply of 78 quarts of drinking water, how many full days will the supply last?
    (a) 103   (b) 80   (c) 79   (d) 59   (e) 11

12. Leianna went to the fish store. She bought 2.3 pound of salmon, 1.125 pounds of halibut and $2\frac{3}{8}$ pounds of flounder. How many pounds of fish did she buy?
    (a) 6   (b) 10.38   (c) 5.8   (d) 6.38   (e) 6.5

13. Ben and Nat shopped for materials for a science project. They need 3 two pound packs of red clay, 2 five pound packs of yellow clay and 3 four pound packs of orange clay. The clay is $1.99 a pound. Approximate the cost of all the clay.
    (a) $12   (b) $40   (c) $16   (d) $25   (e) $56

14. Four friends are shopping for school supplies. They all need the same type of pen that sells for $2.95. They see a pack of 4 such pens for $10.00. If they buy the pack and split the cost, how much money would each of the four students save?
    (a) $0.70   (b) $0.45   (c) $0.50   (d) $2.50   (e) $1.80

15. When 10 friends shared the cost of a ski cabin, the cost per person was $51.60. If 12 friends shared the cabin, how much less would each friend have to pay?

    (a) $4.30   **(b)** $8.60   **(c)** $5.16   **(d)** $3.20   **(e)** $1.80

16. A jacket originally costs $88.00. The price of the jacket is reduced by $\frac{1}{4}$. What is the reduced price?

    (a) $75   **(b)** $60   **(c)** $84   **(d)** $80   **(e)** $66

17. The price of a jacket is reduced by $\frac{1}{5}$ to a sale price of $56.20. What was the original price?

    (a) $70.25   **(b)** $281   **(c)** $67.44   **(d)** $75.00   **(e)** $100

18. A jacket originally costs $80.00. The price is reduced by $\frac{1}{5}$, and then the sale price is further reduced by $\frac{1}{4}$. What is price after both reductions?

    (a) $50   **(b)** $40   **(c)** $55   **(d)** $48   **(e)** $20

19. The price of a jacket is reduced by $\frac{1}{5}$, and then the sale price is further reduced by $\frac{1}{4}$. After both reductions the sale price is $54.00. What was the original price of the jacket?

    (a) $108   **(b)** $90   **(c)** $70   **(d)** $85   **(e)** $74

20. In the number 0.55, the 5 digit on the left represents a value that is how many times as large as the 5 digit on the right?

    (a) $\frac{1}{10}$ **(b)** 50   **(c)** 2   **(d)** 10   **(e)** $\frac{1}{2}$

21. In the number 0.3003, the 3 next to the decimal point represents a value that is how many times as large as the other 3?

    (a) 100   **(b)** 10   **(c)** $\frac{1}{10}$ **(d)** 1000   **(e)** 300

22. Richard earns $2500 dollars a month. He spends $\frac{1}{4}$ of his income on rent and divides the rest evenly between savings and expenses. How much does he save?

    (a) $625   **(b)** $500   **(c)** $312.50   **(d)** $750   **(e)** $937.50

23. A school newspaper has a budget of $2300. The budget goes for printing and supplies. They spend twice as much on printing as on supplies. How much do they spend on supplies?

    (a) $766.67   **(b)** $1533.33   **(c)** $2000   **(d)** $1300   **(e)** $300

24. A taxi ride costs $3.50 for the first $\frac{1}{2}$ mile, and $0.25 for every $\frac{1}{4}$ mile after that. How much will a 10 mile trip cost?

    (a) $30.75   **(b)** $15.25   **(c)** $10.75   **(d)** $13   **(e)** $20

25. The gas company charges a $12.00 monthly "connection fee", an $8.25 monthly "delivery charge", $3.50 for every cubic foot of gas used up to 450 cubic feet, and $2.50 per cubic foot in excess of 450. What is the charge for 500 cubic feet of gas?

    (a) $1720.25   **(b)** $820.25   **(c)** $750   **(d)** $1545.25   **(e)** $450

26. Jim went to the pool hall. There is an entrance fee and a price per game. On Wednesday Jim played one game and paid $5.25 (including the entrance fee). On Thursday the price per game increased and Joe played one game and paid $6.10. By how much did the price per game cost increase?

    (a) $0.35   **(b)** $0.10   **(c)** $1.10   **(d)** $0.85   **(e)** $0.75

27. Jim and Joe went to the pool hall. There is an entrance fee and a price per game. Joe played one game and paid $5.25 (with entrance fee). Jim played five games and paid $8.25 (with entrance fee). What is the cost per game?

    (a) $2.48   **(b)** $0.75   **(c)** $0.50   **(d)** $1.25   **(e)** $1.10

For 28 – 29, a Fahrenheit temperature is 1.8 times the equivalent Celsius temperature plus 32.

28. Find the Fahrenheit temperature when the Celsius temperature is 25 degrees.
    (a) 68°F   **(b)** 70°F   **(c)** 46°F   **(d)** 77°F   **(e)** 52°F

29. Find the Celsius temp when the Fahrenheit temperature is 68 degrees.
    (a) 90°C   **(b)** 5.7°C   **(c)** 15°C   **(d)** 18°C   **(e)** 20°C

30. Steve pays $0.05 per minute for his cell phone calls. In March he was charged $58.95 for calls. How many minutes did he speak?
    (a) 2948   **(b)** 29,475   **(c)** 5895   **(d)** 1179   **(e)** 5900

31. Dale's has 400 free minutes of calls on his cell phone plan, and is charged $0.12 cents per minute for any minutes over 400. In October he was charged $27.00 for calls. How many minutes did he talk?
    (a) 625   **(b)** 500   **(c)** 400   **(d)** 175   **(e)** 350

32. Bob's bill was $56.00. This included an $8.00 service fee. He has 300 free minutes. After the initial 300 minutes he is charged $0.15 a minute. How many minutes did he talk?
    (a) 20   **(b)** 320   **(c)** 840   **(d)** 620   **(e)** 373

For 33 – 34, one Canadian dollar is worth 0.84 U.S. dollars, and one U.S. dollar is worth 12.50 Mexican Pesos.

33. What is one Canadian dollar worth in Mexican Pesos?
    (a) 14.88 Pesos   **(b)** 13.34 Pesos   **(c)** 15 Pesos
    (d) 10.50 Pesos   **(e)** 12 Pesos

34. What are 30 Canadian dollars worth in Mexican Pesos?
    (a) 315 Pesos   **(b)** 120 Pesos   **(c)** 157 Pesos
    (d) 334 Pesos   **(e)** 840 Pesos

Chapter 4

Exponents, Scientific Notation and Square Roots

SECTION A. Positive Powers

An exponent tells us the number of terms a number, such as 10, is to be multiplied by itself. Thus, $3^2 = 3 \times 3$, $4^3 = 4 \times 4 \times 4$, $10^4 = 10 \times 10 \times 10 \times 10$. In general, exponential notation is of the form

$$a^n = \underbrace{a \times a \times a \times \cdots \times a}_{n \text{ factors}}$$

a represents   the exponent
a number

This is usually read: a to the nth power.

In the case where a is 10, we have:

$10^1 = 10$

$10^2 = 10 \times 10 = 100$

$10^3 = 10 \times 10 \times 10 = 1000$

$10^4 = 10 \times 10 \times 10 \times 10 = 10,000$

So when 10 is raised to the nth power, the exponent tells us how many zeros follow the one.

Thus, $10^5 = 100,000$, i.e., 5 zeros follow the one.

When a number is multiplied by $10^1 = 10$, the product is the original number with a zero appended to its right.

For example, $23 \times 10 = 230$

$469 \times 10 = 4690$

Similarly, when a number is multiplied by $10^2 = 100$, the product is the original number with two zeros appended to its right.

For example, $23 \times 100 = 2300$

$469 \times 100 = 46,900$

In general, when a number is multiplied by 10^n, the product is the original number with n zeros appended to its right.

For example, $23 \times 10^4 = 230,000$

$469 \times 10^4 = 4,690,000$

We observe that if the decimal points were present in the number discussed, then our results could easily be obtained by moving the decimal point n positions to the right and filling in blank spaces with zeros.

For example, $23 \times 10 = 23.\underline{0}_{\,} = 230$

$23 \times 10^2 = 23.\underline{00}_{\,} = 2300$

$23 \times 10^3 = 23.\underline{000}_{\,} = 23,000$

$7.96 \times 10^2 = 7.\underline{96}_{\,} = 796$

$0.0967 \times 10^3 = 0.\underline{096}_{\,}7 = 96.7$ ← | Note, we do not write zero to the left of the 9 in the product. |

EXAMPLE 1: Evaluate the following expressions.

   a. $\left(2^2\right)^3$

   b. $(-1)^5$

   c. $\left(\frac{3}{4}\right)^2$

   d. $(-2)^3$

   e. -2^2

SOLUTION:

   a. $\left(2^2\right)^3 = 2^2 \cdot 2^2 \cdot 2^2 = 4 \cdot 4 \cdot 4 = 64$

   b. $(-1)^5 = (-1)(-1)(-1)(-1)(-1) = -1$

   c. $\left(\frac{3}{4}\right)^2 = \frac{3}{4} \cdot \frac{3}{4} = \frac{9}{16}$

   d. $(-2)^3 = (-2)(-2)(-2) = -8$

   e. $-2^2 = -(2)(2) = -4$

EXAMPLE 2: Perform the indicated operation and evaluate.

   a. $2^2 \cdot 8^2$

   b. $\frac{\left(3^2\right)^2}{9}$

   c. $7 \times 10^2 + 8 \times 10^1 + 6$

SOLUTION:

   a. $2^2 \cdot 8^2 = 2 \times 2 \times 8 \times 8 = 256$

   b. $\frac{\left(3^2\right)^2}{9} = \frac{9 \times 9}{9} = 9$

   c. $7 \times 10^2 + 8 \times 10^1 + 6 = 700 + 80 + 6 = 786$

128

Exercises: Positive Powers

For 1 – 5, evaluate each expression.

1. 2^4
    (a) 16   (b) 8   (c) 24   (d) 32   (e) not given

2. $\left(2^3\right)^2$
    (a) 64   (b) 32   (c) 16   (d) 24   (e) not given

3. $(-1)^4$
    (a) 1   (b) –1   (c) 4   (d) –4   (e) not given

4. $\left(\dfrac{2}{3}\right)^3$
    (a) $\dfrac{8}{27}$ (b) $\dfrac{4}{9}$ (c) $\dfrac{6}{3}$ (d) 2   (e) not given

5. -2^4
    (a) 16   (b) –16   (c) 8   (d) –8   (e) not given

For exercises 6 – 12, perform the indicated operations and evaluate.

6. $2^3 \cdot 8^2$
    (a) 32   (b) 128   (c) 320   (d) 480   (e) not given

7. $\dfrac{\left(4^2\right)^3}{16}$
    (a) 1   (b) 256   (c) 64   (d) 1024   (e) not given

8. $2 \cdot 10^3$
    (a) 200   (b) 20   (c) 2000   (d) 800   (e) not given

9. $6 \cdot 10^2$
    (a) 600   (b) 60   (c) 3600   (d) 6000   (e) not given

10. $8 \cdot 10^1$
    (a) 800   **(b)** 80   **(c)** 8000   **(d)** 8   **(e)** not given

11. $2 \cdot 10^3 + 6 \cdot 10^2 + 8 \cdot 10^1 + 7.0$
    (a) 2687   **(b)** 2876   **(c)** 8276   **(d)** 2300   **(e)** not given

12. $8 \cdot 10^2 + 9 \cdot 10^1 + 6.0$
    (a) 23   **(b)** 230   **(c)** 896   **(d)** 968   **(e)** not given

For 13 – 15, find the missing exponent.

13. $8 \times 8 = 2^?$
    (a) 3   **(b)** 9   **(c)** 6   **(d)** 4   **(e)** not given

14. $2^6 \times 8 = 2^?$
    (a) 3   **(b)** 9   **(c)** 6   **(d)** 4   **(e)** not given

15. $8^2 \times 2^6 = 2^?$
    (a) 3   **(b)** 9   **(c)** 6   **(d)** 12   **(e)** not given

For 16 – 20, write each value in regular decimal notation without the power of 10.

16. 4.67×10^2
    (a) 46.7   **(b)** 467   **(c)** 0.467   **(d)** 0.0467   **(e)** not given

17. 467×10^2
    (a) 46.7   **(b)** 46,700   **(c)** 4670   **(d)** 4.67   **(e)** not given

18. 46.7×10^3
    (a) 46.7   **(b)** 46,700   **(c)** 4670   **(d)** 4.67   **(e)** not given

19. 0.467×10^4
    (a) 46.7   **(b)** 46,700   **(c)** 4670   **(d)** 4.67   **(e)** not given

20. 0.00467×10^3
    (a) 46.7   **(b)** 46,700   **(c)** 4670   **(d)** 4.67   **(e)** not given

SECTION B. Negative Powers of 10

When a number is written as the product of a number greater than or equal to 1, and less than 10 and also as a power of 10, it is considered to be in **scientific notation**. For example,

Numbers   Scientific Notation

a. 647 $= 6.47 \times 100 = 6.47 \times 10^2$

b. 64.7 $= 6.47 \times 10 = 6.47 \times 10^1$

c. 6.47 $= 6.47 \times 1 = 6.47$ $\leftarrow$ | Any number greater than or equal to 1 and less than 10 is already written in scientific notation |

d. 64,700 $= 6.47 \times 10,000 = 6.47 \times 10^4$

We can also represent numbers which are less than 1 by using negative powers of 10 where we make use of the notation $10^{-1} = \dfrac{1}{10}$, $10^{-2} = \dfrac{1}{10^2}$, $10^{-3} = \dfrac{1}{10^3}$, etc. to write

$$0.2 = \frac{2}{10} = 2 \times \frac{1}{10} = 2 \times 10^{-1}$$

$$0.02 = \frac{2}{100} = 2 \times \frac{1}{100} = 2 \times 10^{-2}$$

$$0.002 = \frac{2}{1000} = 2 \times \frac{1}{1000} = 2 \times 10^{-3}, \text{ etc.}$$

In general, $10^{-n} = \dfrac{1}{10^n}$.

We notice that multiplying a number by 10^{-n} has the effect of moving the decimal point n positions to the left and filling the blank spaces with zeros.

$$4.6 \times 10^{-1} = .4.6 = 0.46$$

$$4.6 \times 10^{-2} = .04.6 = 0.046$$

$$467 \times 10^{-3} = .467. = 0.467$$

We can now write numbers that are less than 1 in scientific notation. For example,

Numbers	Scientific Notation
a. 0.7	$= 7 \times 10^{-1}$
b. 0.07	$= 7 \times 10^{-2}$
c. 0.007	$= 7 \times 10^{-3}$
d. 0.078	$= 7.8 \times 10^{-2}$
e. 0.000782	$= 7.82 \times 10^{-4}$

Using positive and negative powers of 10, we can write numbers in expanded exponential form:

EXAMPLE 1: Write each of the following in expanded exponential form:
  a. 43
  b. 432
  c. 4328
  d. 4328.1
  e. 4328.12

SOLUTION:

  a. $43 = 40 + 3 = 4 \times 10^1 + 3$

  b. $432 = 400 + 30 + 2 = 4 \times 10^2 + 3 \times 10^1 + 2$

  c. $4328 = 4000 + 300 + 20 + 8 = 4 \times 10^3 + 3 \times 10^2 + 2 \times 10^1 + 8$

  d. $4328.1 = 4000 + 300 + 20 + 8 + 0.1 = 4 \times 10^3 + 3 \times 10^2 + 2 \times 10^1 + 8 + 1 \times 10^{-1}$

  e. $4328.12 = 4000 + 300 + 20 + 8 + 0.1 + 0.02 = 4 \times 10^3 + 3 \times 10^2 + 2 \times 10^1 + 8 + 1 \times 10^{-1} + 2 \times 10^{-2}$

EXAMPLE 2: Write each of the following in decimal notation (without powers of 10):
  a. $4 \times 10^2 + 7 \times 10^1 + 2$
  b. $5 \times 10^2 + 6 \times 10^1 + 5$
  c. $5 \times 10^2 + 6 \times 10^1 + 5 + 7 \times 10^{-1}$
  d. $5 \times 10^2 + 6 \times 10^1 + 5 + 7 \times 10^{-1} + 8 \times 10^{-2}$
  e. $5 + 6 \times 10^{-1} + 7 \times 10^1$
  f. $5 \times 10^{-2} + 7 \times 10^1 + 3$

SOLUTION:

  a. $4 \times 10^2 + 7 \times 10^1 + 2 = 400 + 70 + 2 = 472$

  b. $5 \times 10^2 + 6 \times 10^1 + 5 = 500 + 60 + 5 = 565$

  c. $5 \times 10^2 + 6 \times 10^1 + 5 + 7 \times 10^{-1} = 500 + 60 + 5 + 0.7 = 565.7$

  d. $5 \times 10^2 + 6 \times 10^1 + 5 + 7 \times 10^{-1} + 8 \times 10^{-2} = 500 + 60 + 5 + 0.7 + 0.08 = 565.78$

  e. $5 + 6 \times 10^{-1} + 7 \times 10^1 = 5 + 0.6 + 70 = 75.6$

  f. $5 \times 10^{-2} + 7 \times 10^1 + 3 = 0.05 + 70 + 3 = 73.05$

Exercises: Negative Powers of 10

For 1 – 10, write in scientific notation.

1. 7600

    (a) 7.6×10^3 (b) 0.76×10^3 (c) 7.6×10^4 (d) 7.6×10^2 (e) not given

2. 7.600

    (a) 7.6   (b) 7.60×10^{-1} (c) 0.76×10^2 (d) 7.6×10^2 (e) not given

3. 0.7600

    (a) 0.76   (b) 0.76×10^2 (c) 0.76×10^3 (d) 7.6×10^{-1} (e) not given

4. 0.007600

    (a) 0.0076   (b) 7.6×10^2 (c) 7.6×10^{-6} (d) 7.6×10^{-3} (e) not given

5. 0.07162

    (a) 0.071×10^2 (b) 0.0716×10^{-2} (c) 0.07612
    (d) 7.612×10^{-2} (e) not given

6. 76.0×10^3

    (a) 7.6×10^5 (b) 0.76×10^3 (c) 7.6   (d) 76,000   (e) not given

7. 760.00×10^8

    (a) 7.6×10^{10} (b) 7.6×10^6 (c) 7.6×10^{-2} (d) 0.76×10^{12} (e) not given

8. 0.7612×10^{14}

    (a) 7.6×10^8 (b) 7.6×10^{10} (c) 7.612×10^{-13}
    (d) 7.612×10^{15} (e) not given

9. 0.07612×10^{-8}

    (a) 7.612×10^{-6} (b) 7.612×10^{-10} (c) 7.6×10^{-10}
    (d) 0.07612×10^{-6} (e) not given

133

10. 7,612,000,000
    (a) 7.612×10^4 **(b)** 7.612×10^{-9} **(c)** 7.612×10^9
    (d) 7.612×10^{11} **(e)** not given

For exercises 11 – 18, write each number in decimal notation (without powers of 10).

11. 5.008×10^{-3}
    (a) 0.005008   **(b)** 0.05008   **(c)** 0.5008   **(d)** 5.008   **(e)** 5008.0

12. 0.00508×10^{-1}
    (a) 0.005080   **(b)** 0.000508   **(c)** 0.5008   **(d)** 0.0508   **(e)** 0.5080

13. 2.37×10^{-3}
    (a) 0.00237   **(b)** 2370   **(c)** 23.70   **(d)** 0.2370   **(e)** 237.00

14. $2\times10^2+3\times10^1+4+7\times10^{-1}+2\times10^{-2}$
    (a) 234.72   **(b)** 23.472   **(c)** 2347.2   **(d)** 23.472   **(e)** not given

15. $2.0+3\times10^{-1}+4\times10^1$
    (a) 23.4   **(b)** 32.4   **(c)** 42.3   **(d)** 43.2   **(e)** 432

16. $5\times10^{-3}+6.0+6.0\times10^{-2}+7.0\times10^{-1}$
    (a) 67.65   **(b)** 6.765   **(c)** 6765   **(d)** 0.6765   **(e)** not given

17. $5.0\times10^3+4.0+7.0\times10^{-2}$
    (a) 5.0047   **(b)** 547.4   **(c)** 5474   **(d)** 5004.07   **(e)** 5004.7

18. $(5\times10^{-4})+(2.0\times10^{-1})+(6.0)+(3\times10^2)$
    (a) 306.2005   **(b)** 306.25   **(c)** 300.605   **(d)** 3602.3   **(e)** not given

For exercises 19 – 21, find the value of *n*.

19. $0.0000387 = 3.87 \times 10^n$
   (a) −4   **(b)** −5   **(c)** −3   **(d)** 2   **(e)** 1

20. $189.67 = 18.967 \times 10^n$
   (a) −4   **(b)** −5   **(c)** −3   **(d)** 2   **(e)** 1

21. $0.00907 = 9.07 \times 10^n$
   (a) −4   **(b)** −5   **(c)** −3   **(d)** 2   **(e)** 1

For exercises 22 – 24, how many zeros will each of the following have when the numbers are written in regular decimal form (without powers of 10)?

22. 126×10^5
   (a) 8   **(b)** 7   **(c)** 6   **(d)** 5   **(e)** 4

23. 19.67×10^7
   (a) 8   **(b)** 7   **(c)** 6   **(d)** 5   **(e)** 4

24. 196.7×10^9
   (a) 8   **(b)** 7   **(c)** 6   **(d)** 5   **(e)** 4

25. The mass of a proton is 0.00000000000000000000000000167 kilogram. Write this in scientific notation.
   (a) 1.67×10^{-27} **(b)** 0.167×10^{-25} **(c)** 1.67×10^{21} **(d)** 1.67×10^{-24} **(e)** not given

26. A single human red blood cell is about 7×10^{-6} meter in diameter. Write this in decimal notation.
   (a) 0.0067   **(b)** 0.00076   **(c)** 0.000007   **(d)** 0.0000007   **(e)** not given

135

SECTION C. Arithmetic Operations Using Scientific Notation

If we try to multiply and divide powers of 10, we will see some interesting patterns which can be generalized into rules.

$$10^3 \times 10^2 = (10 \times 10 \times 10) \times (10 \times 10) = 10^5 = 10^{3+2}$$

$$10^3 \times 10^5 = (10 \times 10 \times 10) \times (10 \times 10 \times 10 \times 10 \times 10) = 10^8 = 10^{3+5}$$

EXAMPLE 1:   Write the following in powers of 10:

    a. $10^2 \times 10^4$

    b. $10^1 \times 10^3$

    c. $10^1 \times 10^6$

    d. $10^2 \times 10^7$

SOLUTION:

    a. $10^2 \times 10^4 = 10^{2+4} = 10^6$

    b. $10^1 \times 10^3 = 10^{1+3} = 10^4$

    c. $10^1 \times 10^6 = 10^{1+6} = 10^7$

    d. $10^2 \times 10^7 = 10^{2+7} = 10^9$

We notice that when we multiply powers of 10, we need only add the exponents. Thus, $10^2 \times 10^7 = 10^{2+7} = 10^9$.

Similarly, with negative exponents:

$$10^{-2} \times 10^{-3} = \frac{1}{10^2} \times \frac{1}{10^3} = \frac{1}{10^{2+3}} = \frac{1}{10^5} = 10^{-5} = 10^{(-2)+(-3)}$$

EXAMPLE 2:   Write the following in powers of 10:

    a. $10^{-2} \times 10^{-4}$

    b. $10^{-1} \times 10^{-3}$

    c. $10^{-2} \times 10^{-7}$

SOLUTION:

    a. $10^{-2} \times 10^{-4} = 10^{-2+(-4)} = 10^{-6}$

    b. $10^{-1} \times 10^{-3} = 10^{-1+(-3)} = 10^{-4}$

    c. $10^{-2} \times 10^{-7} = 10^{-2+(-7)} = 10^{-9}$

Even with both positive and negative exponents combined,

$$10^5 \times 10^{-3} = 10 \times 10 \times 10 \times 10 \times 10 \times \frac{1}{10^3} = \frac{10 \times 10 \times 10 \times 10 \times 10}{10 \times 10 \times 10} = 10 \times 10 = 10^2 = 10^{5+(-3)}$$

$$10^3 \times 10^{-3} = 10 \times 10 \times 10 \times \frac{1}{10^3} = \frac{10 \times 10 \times 10}{10 \times 10 \times 10} = 1 = 10^{3+(-3)}$$

Note, from the last example, $1 = 10^{3+(-3)} = 10^0$, so we define 10^0 to be 1.

EXAMPLE 3: Write the following in powers of 10:

    a. $10^3 \times 10^{-2}$

    b. $10^7 \times 10^{-5}$

    c. $10^4 \times 10^{-4}$

    d. $10^{-4} \times 10^{-5}$

    e. $10^9 \times 10^{-6}$

SOLUTION:

    a. $10^3 \times 10^{-2} = 10^{3+(-2)} = 10^1$

    b. $10^7 \times 10^{-5} = 10^{7+(-5)} = 10^2$

    c. $10^4 \times 10^{-4} = 10^{4+(-4)} = 10^0 = 1$

    d. $10^{-4} \times 10^{-5} = 10^{-4+(-5)} = 10^{-9}$

    e. $10^9 \times 10^{-6} = 10^{9+(-6)} = 10^3$

Now, let's divide using powers of 10:

$$\frac{10^5}{10^3} = \frac{10 \times 10 \times 10 \times 10 \times 10}{10 \times 10 \times 10} = 10 \times 10 = 10^2 = 10^{5-3}$$

$$\frac{10^{-2}}{10^{-3}} = \frac{\frac{1}{10^2}}{\frac{1}{10^3}} = \frac{1}{10^2} \times \frac{10^3}{1} = \frac{10^3}{10^2} = \frac{10 \times 10 \times 10}{10 \times 10} = 10^1 = 10^{-2-(-3)}$$

$$\frac{10^5}{10^{-3}} = \frac{10^5}{\frac{1}{10^3}} = 10^5 \times \frac{10^3}{1} = 10^8 = 10^{5-(-3)}$$

Thus, $\dfrac{10^m}{10^n} = 10^{m-n}$.

In general, when we powers of 10, we simply subtract the exponent of the denominator from the exponent of the numerator. This rule works for all exponents.

EXAMPLE 4: Write the following in powers of 10:

  a. $10^2 \div 10^1$

  b. $10^4 \div 10^3$

  c. $10^{-7} \div 10^3$

  d. $\dfrac{10^5}{10^3}$

  e. $\dfrac{10^6}{10^{-2}}$

  f. $10^7 \div 10^{-2}$

  g. $10^{-7} \div 10^{-2}$

SOLUTION:

  a. $10^2 \div 10^1 = \dfrac{10^2}{10^1} = 10^{2-1} = 10^1$

  b. $10^4 \div 10^3 = \dfrac{10^4}{10^3} = 10^{4-3} = 10^1$

  c. $10^{-7} \div 10^3 = \dfrac{10^{-7}}{10^3} = 10^{-7-3} = 10^{-10}$

  d. $\dfrac{10^5}{10^3} = 10^{5-3} = 10^2$

  e. $\dfrac{10^6}{10^{-2}} = 10^{6-(-2)} = 10^8$

  f. $10^7 \div 10^{-2} = \dfrac{10^7}{10^{-2}} = 10^{7-(-2)} = 10^9$

  g. $10^{-7} \div 10^{-2} = \dfrac{10^{-7}}{10^{-2}} = 10^{-7-(-2)} = 10^{-5}$

When numbers are in scientific notation, it is fairly simple to perform arithmetic operations on them:

$$\frac{2.4 \times 10^7}{1.2 \times 10^4} = \frac{2.4}{1.2} \times \frac{10^7}{10^4} = 2.0 \times 10^3$$

$$\left(1.2 \times 10^3\right) \cdot \left(1.2 \times 10^5\right) = (1.2)(1.2) \times 10^3 \times 10^5 = 1.44 \times 10^8$$

$$\frac{2.88 \times 10^{12}}{\left(1.2 \times 10^3\right) \cdot \left(1.2 \times 10^5\right)} = \frac{2.88 \times 10^{12}}{1.44 \times 10^8} = \frac{2.88}{1.44} \times \frac{10^{12}}{10^8} = 2.0 \times 10^4$$

The best strategy is to deal with the decimal parts separately and the power of 10 separately.

When dealing with arithmetic operations on very large or very small numbers, the best thing to do is to convert these numbers into scientific notation and then perform the arithmetic.

EXAMPLE 5: $(160,000) \times (0.00000003)$

SOLUTION: $(1.6 \times 10^5) \times (3.0 \times 10^{-8}) = (1.6)(3.0) \times 10^5 \times 10^{-8} = 4.8 \times 10^{-3} = 0.0048$

EXAMPLE 6: $\dfrac{8,000,000 \times 0.00008}{200,000 \times 0.0004}$

SOLUTION: $\dfrac{(8 \times 10^6) \cdot (8 \times 10^{-5})}{(2 \times 10^5) \cdot (4 \times 10^{-4})} = \dfrac{8 \times 8}{2 \times 4} \times \dfrac{10^6 \times 10^{-5}}{10^5 \times 10^{-4}} = 8 \times 10^0 = 8.0$

EXAMPLE 7: The average distance of Earth to the sun is almost 93,000,000 miles. If light travels at a speed of 186,000 miles per second, how long does it take light to travel from the sun to Earth?

SOLUTION: $\dfrac{93,000,000}{186,000} = \dfrac{9.3 \times 10^7}{1.86 \times 10^5} = 5 \times 10^2 = 500$ seconds or $\dfrac{500}{60} = 8\dfrac{1}{3}$ minutes!

139

Exercises: Arithmetic Operations Using Scientific Notation

For 1 – 13, perform the indicated operations and write the result in scientific notation.

1. $\dfrac{18.6 \times 10^3}{9.3 \times 10^4}$

   (a) 2×10^1 **(b)** 2×10^{-1} **(c)** 2×10^7 **(d)** 0.5×10^{-7} **(e)** not given

2. $\dfrac{(2 \times 10^3) \times (9.3 \times 10^5)}{18.6 \times 10^{11}}$

   (a) 1×10^3 **(b)** 2×10^{-3} **(c)** 1×10^{-3} **(d)** 1×10^1 **(e)** not given

3. $\dfrac{(1.5 \times 10^2) \times (1.5 \times 10^7)}{(2.25 \times 10^3) \times (2.5 \times 10^{-4})}$

   (a) 4×10^9 **(b)** 1×10^{10} **(c)** 4×10^{10} **(d)** 1.5×10^9 **(e)** not given

4. $\dfrac{(1.6 \times 10^8) \times (4.0 \times 10^2)}{(4 \times 10^2) \times (4 \times 10^3) \times (4 \times 10^{-3})}$

   (a) 1.0×10^8 **(b)** 1×10^7 **(c)** 1.6×10^8 **(d)** 1.6×10^7 **(e)** not given

5. $\dfrac{(3.2 \times 10^{-4}) \times (3.0 \times 10^{-5})}{9.6 \times 10^{-8}}$

   (a) 1×10^{-2} **(b)** 1×10^{-1} **(c)** 3×10^{-2} **(d)** 3×10^{-1} **(e)** 1×10^{-3}

6. $\dfrac{3.3 \times 10^7}{1,100}$

   (a) 1×10^5 **(b)** 3×10^5 **(c)** 3×10^4 **(d)** 3×10^3 **(e)** 3×10^6

7. $\dfrac{6.6 \times 10^{-8}}{1.1}$

   (a) 6×10^{-8} **(b)** 6×10^{-7} **(c)** 6×10^{-9} **(d)** 5.5×10^{-8} **(e)** not given

140

8. $\left(3 \times 10^4\right) \cdot (1100)$

    (a) 3.3×10^7 **(b)** 33×10^6 **(c)** 3.3×10^7 **(d)** 3.3×10^6 **(e)** not given

9. $(2.1)(300,000)$

    (a) 6.3×10^6 **(b)** 6.3×10^5 **(c)** 6.3×10 **(d)** 2.1×10^5 **(e)** not given

10. $\dfrac{\left(2 \times 10^6\right) \times \left(3.3 \times 10^{-2}\right)}{1.1 \times 10^{10}}$

    (a) 6×10^4 **(b)** 6.3×10^5 **(c)** 6×10^{-5} **(d)** 6×10^{-7} **(e)** not given

11. $\dfrac{\left(1.672 \times 10^{-23}\right)}{0.8 \times 10^2}$

    (a) 2.09×10^{-25} **(b)** 2.09×10^{-24} **(c)** 1×10^{-21} **(d)** 2.09×10^{-21} **(e)** 2.09×10^{-24}

12. $\left(5.3 \times 10^{-23}\right) \cdot (1,000,000)$

    (a) 5.3×10^{-29} **(b)** 5.3×10^{-17} **(c)** 1×10^{-17} **(d)** 1×10^{-29} **(e)** not given

13. $\dfrac{5.3 \times 10^{-23}}{1,000,000}$

    (a) 5.3×10^{-29} **(b)** 5.3×10^{-17} **(c)** 1×10^{-17} **(d)** 1×10^{-29} **(e)** not given

14. A spaceship took 310 days to travel 520,000,000 miles. What was the approximate number of miles it averaged per day?

    (a) 1.2×10^7 **(b)** 1.3×10^7 **(c)** 1.7×10^6 **(d)** 1.6×10^6 **(e)** 1.7×10^7

15. If a carbon atom weighs 5.3×10^{-23} grams, how much does 1 million carbon atoms weigh?

    (a) 5.3×10^{-29} **(b)** 5.3×10^{-32} **(c)** 5.3×10^{-10} **(d)** 5.3×10^{-21} **(e)** 5.3×10^{-17}

16. The mass of a neutron is approximately 1.675×10^{-27} kilogram. Find the mass of 180,000 neutrons.

    (a) 3.015×10^{-22} **(b)** 3×10^{-25} **(c)** 3.15×10^{-20}

    (d) 3.015×10^{-20} **(e)** not given

17. The sun radiates into space at a rate of 3.9×10^{26} joules per second. How many joules are emitted in two million seconds?

    (a) 7.8×10^{32} **(b)** 7.8×10^{20} **(c)** 7.8×10^{12} **(d)** 3.9×10^{30} **(e)** not given

18. Avogadro's number says that there are approximately 6.02×10^{23} molecules/mole. About how many molecules can one expect in 0.00483 mole?

    (a) 29×10^{26} **(b)** 29×10^{25} **(c)** 2.9×10^{20} **(d)** 2.9×10^{21} **(e)** not given

For exercises 19 and 20, use the following information: For fiscal year 1999, the national debt was determined to be approximately 5.614×10^{12} dollars.

19. The census bureau estimates that in 1999, the entire population of the United States was 2.76×10^{8} people. If the national debt were evenly divided among every person in the country, how much debt would be assigned to each individual? Round to two decimal places.

    (a) 2.03×10^{3} **(b)** 2.03×10^{4} **(c)** 2.03×10^{2} **(d)** 2.03×10^{5} **(e)** 2.03×10^{1}

20. The census bureau estimates that in 1999, the number of people in the United States who were over age 18 was approximately 2.05×10^{8} people. If the national debt were evenly divided among every person over age 18 in the country, how much debt would be assigned to each individual? Round to two decimal places.

    (a) 2.73×10^{3} **(b)** 2.73×10^{4} **(c)** 2.73×10^{2} **(d)** 2.73×10^{5} **(e)** 2.73×10^{1}

SECTION D. Square Roots and Perfect Squares

The square root of a number is a number that we must square to get the given number. Since $2^2 = 4$, the square root of 4 is 2. Another square root of 4 is –2 since $(-2)^2 = 4$.

The symbol $\sqrt{}$ is used in mathematics for the **principal square root** of a number, which is the nonnegative square root. The symbol itself $\sqrt{}$ is called the **radical sign**. If we want to find the negative square root of a number, we use the symbol $-\sqrt{}$

Definition of Principal Square Root

For all nonnegative numbers N, the principal square root of N (written $\sqrt{N}$) is defined to be the nonnegative number a if and only if $a^2 = N$.

Notice in the definition that we did not use the words "for all *positive* numbers N" because we also want to include the square root of 0. $\sqrt{0} = 0$ since $0^2 = 0$.

For example,

    a. $\sqrt{9} = 3$ because 3 is nonnegative and $3^2 = 9$.

    b. $-\sqrt{9} = -3$ i.e., –3 is the negative square root of 9.

    c. $\sqrt{36} = 6$ because 6 is nonnegative and $6^2 = 36$.

    d. $\sqrt{0.9} = 0.3$ because 0.3 is nonnegative and $(0.3)^2 = 0.09$.

    e. $\sqrt{1600} = 40$ because 40 is nonnegative and $40^2 = 1600$.

The number beneath the radical sign is called the **radicand**. It should be noted that when we deal with square roots, the radicand must be nonnegative. $\sqrt{-9}$ does not exist as a real number, since there is not a real number whose square is –9.

When a number is squared, the resulting number is called a **perfect square**. The number 16 is a perfect square, because $4^2 = 16$. However, 15 is not a perfect square. There is no number that is represented as an integer or fraction or decimal that when squared yields 15. The square roots of a perfect square is an exact number. It is helpful to know the first 15 perfect squares for whole numbers:

$\underbrace{\text{perfect squares}}$
↓

$1^2 = 1$	$\sqrt{1} = 1$
$2^2 = 4$	$\sqrt{4} = 2$
$3^2 = 9$	$\sqrt{9} = 3$
$4^2 = 16$	$\sqrt{16} = 4$
$5^2 = 25$	$\sqrt{25} = 5$

$\underbrace{\text{perfect squares}}$
↓

$6^2 = 36$	$\sqrt{36} = 6$
$7^2 = 49$	$\sqrt{49} = 7$
$8^2 = 64$	$\sqrt{64} = 8$
$9^2 = 81$	$\sqrt{81} = 9$
$10^2 = 100$	$\sqrt{100} = 10$

$\underbrace{\text{perfect squares}}$
↓

$11^2 = 121$	$\sqrt{121} = 11$
$12^2 = 144$	$\sqrt{144} = 12$
$13^2 = 169$	$\sqrt{169} = 13$
$14^2 = 196$	$\sqrt{196} = 14$
$15^2 = 225$	$\sqrt{225} = 15$

For example,

a. 30 is not a perfect square. There is no number that when squared equals 30.

b. 64 is a perfect square because $8^2 = 8 \cdot 8 = 64$

c. $\frac{1}{4}$ is a perfect square because $\left(\frac{1}{2}\right)^2 = \frac{1}{2} \cdot \frac{1}{2} = \frac{1}{4}$

144

SECTION E. Approximating Square Roots

If the radicand is not a perfect square, it is impossible to represent the square root exactly as a terminating or repeating decimal. For example, $\sqrt{2}$ cannot be represented exactly as a decimal, because there is no decimal number whose square is 2. However, we can approximate $\sqrt{2}$ very closely, by using as many decimal places desired. We observe

$$(1.4)^2 = 1.96$$
$$\left(\sqrt{2}\right)^2 = 2$$
$$(1.5)^2 = 2.25$$

Since 2 is between 1.96 and 2.25, $\sqrt{2}$ is between 1.4 and 1.5, closer to 1.4 because 1.96 is closer to 2 than is 2.25.

$$(1.41)^2 = 1.9881$$
$$\left(\sqrt{2}\right)^2 = 2$$
$$(1.42)^2 = 2.0162$$

Since 2 is between 1.9881 and 2.0162, $\sqrt{2}$ is between 1.41 and 1.42, closer to 1.41 because 1.9881 is closer to 2 than is 2.0162. We can continue in this way and get better and better approximations to $\sqrt{2}$, but we could never get the exact value. Thus:

to the nearest tenth $\sqrt{2} \approx 1.4$,

to the nearest hundredth $\sqrt{2} \approx 1.41$, $\approx$ means approximately equal

to the nearest thousandth $\sqrt{2} \approx 1.414$.

Numbers such as $\sqrt{2}$ are called **irrational** numbers. These numbers cannot be represented by terminating or repeating decimals, but these numbers do exist even if we can't represent them! Only rational numbers can be represented by terminating or repeating decimals.

EXAMPLE 1: Identify the following numbers as rational or irrational. If rational, find the square root exactly. If irrational, find two consecutive whole numbers for which the exact value is between.

    a. $\sqrt{121}$

    b. $\sqrt{12}$

    c. $\sqrt{2.25}$

    d. $\sqrt{250}$

SOLUTION:

    a. $\sqrt{121}$ is rational since $\sqrt{121} = 11$.

    b. $\sqrt{12}$ is irrational. $3^2 = 9$, $4^2 = 16$. Since 12 is between 9 and 16, $\sqrt{12}$ is between 3 and 4

    c. $\sqrt{2.25}$ is rational since $\sqrt{2.25} = 1.5$.

    d. $\sqrt{250}$ is irrational. $20^2 = 400$ is much too huge. Try $15^2 = 225$, a bit too low. Try $(16)^2 = 256$, a bit too high. So $\sqrt{250}$ is between 15 and 16.

EXAMPLE 2: $\sqrt{3000}$ is closest to which integer?

SOLUTION: Try $(10)^2 = 100$, much too low. Next try $(50)^2 = 2500$, still too low. Try $(60)^2 = 3600$, too high. Try between 50 and 60, $(55)^2 = 3025$, still a bit too high. Try $(54)^2 = 2916$, now a little too low. Thus $\sqrt{3000}$ is between 54 and 55, but closer to 55 since 3025 is closer to 3000 than is 2916.

EXAMPLE 3: We know that the area of a square with side of length s is s^2, i.e. if we know s^2, we can get the side s by simply finding the square root of s^2 $\left(\sqrt{s^2} = s\right)$. If a square field is 43,000 square feet, what is the length of one side of the square field?

SOLUTION: We have to find $\sqrt{43,000}$. First try $(200)^2 = 40,000$, so 200 is a bit too low. Try $(300)^2 = 90,000$, so 300 is much too high. So try $(210)^2 = 44,100$, a drop too high. Try $(205)^2 = 42,025$, too low. And $(207)^2 = 42,849$, still too low. Next try $(208)^2 = 43,264$. Thus, $\sqrt{43,000}$ is between 207 and 208, but closer to 207 because 43,000 is closer to 42,849 than to 43,264.

146

Exercises: Approximating Square Roots

For 1 – 4, find the two square roots of each number.

1. 9   2. 4   3. 49   4. 36

For 5 – 22, find the square root.

5. $\sqrt{36}$ 6. $\sqrt{25}$ 7. $\sqrt{81}$ 8. $\sqrt{64}$

9. $-\sqrt{49}$ 10. $-\sqrt{16}$ 11. $\sqrt{0.81}$ 12. $\sqrt{0.64}$

13. $\sqrt{900}$ 14. $\sqrt{2500}$ 15. $-\sqrt{10,000}$ 16. $\sqrt{40,000}$

17. $\sqrt{144}$ 18. $\sqrt{256}$ 19. $\sqrt{0.0049}$ 20. $\sqrt{0.0025}$

21. $\sqrt{16,900}$ 22. $\sqrt{22,500}$

23. A tree is 12 feet tall. A wire that stretches from a point that is 5 feet from the base of the tree to the top of the tree is $\sqrt{169}$ feet long. Exactly how many feet long is the wire?
    (a) 12.5   (b) 12.6   (c) 13   (d) 14   (e) 15

24. A tree is 24 feet tall. A wire that stretches from a point that is 10 feet from the base of the tree to the top of the tree is $\sqrt{676}$ feet long. Exactly how many feet long is the wire?
    (a) 24   (b) 25   (c) 26   (d) 27   (e) 36

For 25 – 32, find the following square roots. If the square root does not exist as a real number, state so.

25. $\sqrt{36}$ 26. $-\sqrt{49}$ 27. $\sqrt{-36}$ 28. $\sqrt{64}$

29. $\sqrt{-64}$ 30. $\sqrt{1}$ 31. $\sqrt{-1}$ 32. $-\sqrt{1}$

147

For 33 – 40, for the following irrational numbers, find the integer that is closest to it.

33. $\sqrt{3900}$
(a) 60   (b) 61   (c) 62   (d) 63   (e) 70

34. $\sqrt{245}$
(a) 14   (b) 15   (c) 16   (d) 17   (e) 20

35. $\sqrt{500}$
(a) 21   (b) 22   (c) 23   (d) 24   (e) 50

36. $\sqrt{180}$
(a) 13   (b) 14   (c) 15   (d) 16   (e) 17

37. $\sqrt{320}$
(a) 16   (b) 17   (c) 18   (d) 19   (e) 20

38. $\sqrt{680}$
(a) 25   (b) 26   (c) 27   (d) 28   (e) 29

39. $\sqrt{1280}$
(a) 35   (b) 36   (c) 37   (d) 38   (e) 39

40. $\sqrt{800}$
(a) 27   (b) 28   (c) 29   (d) 30   (e) 31

For 41 – 43, find the approximate length of the side of a square whose area is

41. 250,000 square feet
(a) 490   (b) 500   (c) 510   (d) 520   (e) 525

42. 90,000 square feet
(a) 290   (b) 300   (c) 310   (d) 320   (e) 330

43. 245 square feet
(a) 25.3   (b) 15.7   (c) 16.2   (d) 17.1   (e) 18.2

SECTION F. Simplifying Square Roots

We note $\sqrt{144} = \sqrt{9 \cdot 16} = \sqrt{9} \cdot \sqrt{16} = 3 \cdot 4 = 12$ and $\sqrt{144} = 12$. Thus, if we first factor the radicand and then take the square root of each factor or we directly take the square root of the radicand without factoring we arrive at the same result. We generalize and have the following rule.

Rule 1: $\sqrt{ab} = \sqrt{a} \cdot \sqrt{b}$ The square root of a product equals the product of the square roots.

Similarly, for division, the square root of a quotient, we can find the square root of the numerator and divide by the square root of the denominator.

Rule 2: $\sqrt{\dfrac{a}{b}} = \dfrac{\sqrt{a}}{\sqrt{b}}$ The square root of a quotient equals the quotient of the square roots.

EXAMPLE 1:

   a. $\sqrt{2500} = \sqrt{25 \times 100} = \sqrt{25} \cdot \sqrt{100} = 5 \times 10 = 50$

   b. $\sqrt{0.9} = \sqrt{\dfrac{9}{10}} = \dfrac{\sqrt{9}}{\sqrt{10}} = \dfrac{3}{10} = 0.3$

   c. $\sqrt{\dfrac{4}{9}} = \dfrac{\sqrt{4}}{\sqrt{9}} = \dfrac{2}{3}$

   d. $\sqrt{\dfrac{121}{16}} = \dfrac{\sqrt{121}}{\sqrt{16}} = \dfrac{11}{4}$

   e. $\sqrt{\dfrac{36}{49}} = \dfrac{\sqrt{36}}{\sqrt{49}} = \dfrac{6}{7}$

   f. $\sqrt{250,000} = \sqrt{25 \times 10,000} = \sqrt{25} \cdot \sqrt{10,000} = 5 \times 100 = 500$

   g. $\sqrt{90,000} = \sqrt{9 \times 10,000} = \sqrt{9} \cdot \sqrt{10,000} = 3 \times 100 = 300$

When we have a product of a square root and a number such as $3\sqrt{5}$, we write the number (coefficient of the square root) on the left side of the square root. Thus, $3\sqrt{5}$ and not $\sqrt{5} \cdot 3$ (3 is the coefficient). $\uparrow$ coefficient

   h. $\sqrt{27} = \sqrt{9 \times 3} = \sqrt{9} \times \sqrt{3} = 3\sqrt{3}$

   i. $\sqrt{18} = \sqrt{9 \times 2} = \sqrt{9} \times \sqrt{2} = 3\sqrt{2}$

   j. $\sqrt{500} = \sqrt{5 \times 100} = \sqrt{5} \times \sqrt{100} = (\sqrt{5})(10) = 10\sqrt{5}$

149

In some instances, it is useful to use Rule 1 and Rule 2 in reverse order, i.e. $\sqrt{a} \cdot \sqrt{b} = \sqrt{ab}$ and $\dfrac{\sqrt{a}}{\sqrt{b}} = \sqrt{\dfrac{a}{b}}$.

EXAMPLE 2:

a. $\sqrt{5} \cdot \sqrt{5} = \sqrt{5 \times 5} = \sqrt{25} = 5$ Note: in general, $\sqrt{a} \cdot \sqrt{a} = \left(\sqrt{a}\right)^2 = a$

b. $\sqrt{8} \cdot \sqrt{2} = \sqrt{8 \times 2} = \sqrt{16} = 4$

c. $\dfrac{\sqrt{12}}{\sqrt{3}} = \sqrt{\dfrac{12}{3}} = \sqrt{4} = 2$

d. $\dfrac{\sqrt{96}}{\sqrt{6}} = \sqrt{\dfrac{96}{6}} = \sqrt{16} = 4$

e. $\sqrt{20} \cdot \sqrt{5} = \sqrt{20 \times 5} = \sqrt{100} = 10$

f. $\left(3\sqrt{5}\right)^2 = \left(\sqrt{9} \cdot \sqrt{5}\right)^2 = \left(\sqrt{45}\right)^2 = 45$

g. $\left(3\sqrt{5}\right)^2 = \left(3\sqrt{5}\right)\left(3\sqrt{5}\right) = (3 \cdot 3)\left(\sqrt{5} \cdot \sqrt{5}\right) = 9 \cdot 5 = 45$

h. $\left(3\sqrt{8}\right)\left(5\sqrt{2}\right) = (3 \cdot 5)\left(\sqrt{8} \cdot \sqrt{2}\right) = 15\sqrt{16} = 15(4) = 60$

Note: $\left.\begin{array}{l} \sqrt{16+9} = \sqrt{25} = 5 \\ \sqrt{16} + \sqrt{9} = 4 + 3 = 7 \end{array}\right\} \sqrt{16+9} \neq \sqrt{16} + \sqrt{9}$

In general, $\sqrt{a+b} \neq \sqrt{a} + \sqrt{b}$. Similarly, $\sqrt{a-b} \neq \sqrt{a} - \sqrt{b}$.

EXAMPLE 3:

a. $\sqrt{25-16} = \sqrt{9} = 3$, but $\sqrt{25-16} \neq \sqrt{25} - \sqrt{16} = 5 - 4 = 1$

b. $\sqrt{100-64} = \sqrt{36} = 6$

c. $\sqrt{100} - \sqrt{64} = 10 - 8 = 2$

d. $\sqrt{225-144} = \sqrt{81} = 9$

e. $\sqrt{144+81} = \sqrt{225} = 15$

f. $\sqrt{81} - \sqrt{16} = 9 - 4 = 5$

g. $\sqrt{25} + \sqrt{16} = 5 + 4 = 9$

Exercises: Simplifying Square Roots

For 1 – 20, evaluate.

1. $\sqrt{3} \times \sqrt{12}$
   (a) 36   (b) 6   (c) 3   (d) 12   (e) not given

2. $\sqrt{18} \times \sqrt{2}$
   (a) 36   (b) 6   (c) 3   (d) 12   (e) not given

3. $\sqrt{32} \times \sqrt{2}$
   (a) 32   (b) 3   (c) 2   (d) 8   (e) not given

4. $\dfrac{\sqrt{32}}{\sqrt{2}}$
   (a) 16   (b) 32   (c) 4   (d) 8   (e) not given

5. $\dfrac{\sqrt{72}}{\sqrt{18}}$
   (a) 4   (b) 2   (c) 8   (d) 6   (e) not given

6. $\dfrac{\sqrt{27}}{\sqrt{3}}$
   (a) 9   (b) 27   (c) 9   (d) 2   (e) not given

7. $\left(4\sqrt{3}\right)^2$
   (a) 16   (b) 24   (c) 48   (d) 8   (e) not given

8. $\left(2\sqrt{5}\right)\left(3\sqrt{5}\right)$
   (a) 6   (b) 5   (c) 30   (d) 12   (e) not given

9. $\left(3\sqrt{12}\right)\left(2\sqrt{3}\right)$
   (a) 6   (b) 36   (c) 10   (d) 8   (e) not given

10. $\sqrt{16} + \sqrt{9}$
    (a) 7   (b) 5   (c) $\sqrt{25}$ (d) 8   (e) not given

11. $\sqrt{16+9}$

    (a) 7   **(b)** 5   **(c)** 6   **(d)** 8   **(e)** not given

12. $\sqrt{100-36}$

    (a) 4   **(b)** 8   **(c)** 6   **(d)** 7   **(e)** not given

13. $\sqrt{100}-\sqrt{36}$

    (a) 4   **(b)** 8   **(c)** 6   **(d)** 7   **(e)** not given

14. $\sqrt{225}-\sqrt{81}$

    (a) 12   **(b)** 144   **(c)** 15   **(d)** 14   **(e)** not given

15. $\sqrt{250,000}-\sqrt{90,000}$

    (a) 200   **(b)** 500   **(c)** 400   **(d)** $\sqrt{160,000}$ **(e)** not given

16. $\sqrt{36+64}$

    (a) 100   **(b)** 10   **(c)** 14   **(d)** 12   **(e)** not given

17. $\dfrac{\sqrt{100-64}}{\sqrt{36}}$

    (a) 6   **(b)** 1   **(c)** 8   **(d)** $\dfrac{2}{6}$ **(e)** not given

18. $\dfrac{\sqrt{36+64}}{\sqrt{25}}$

    (a) 10   **(b)** 4   **(c)** 2   **(d)** 3   **(e)** not given

19. $\dfrac{\sqrt{4\times16}}{\sqrt{9+16}}$

    (a) $\dfrac{8}{3}$ **(b)** $\dfrac{8}{5}$ **(c)** 4   **(d)** 2   **(e)** not given

20. $\dfrac{\sqrt{4\times16}}{\sqrt{2\times16}}$

    (a) 4   **(b)** 8   **(c)** 2   **(d)** 1   **(e)** not given

SECTION G. Adding and Subtracting Like Square Roots

Square roots with the same radicands are called like square roots. The following pairs are like square roots: $\sqrt{3}$ and $2\sqrt{3}$, $\sqrt{5}$ and $5\sqrt{5}$, and $3\sqrt{12}$ and $5\sqrt{12}$. We can add or subtract like square roots only. We cannot add or subtract unlike square roots, such as $\sqrt{3}+\sqrt{2}$.

To add or subtract like square roots, add or subtract the coefficients and leave the square root portions unchanged.

For example,
   a. $7\sqrt{2}+5\sqrt{2}=12\sqrt{2}$
   b. $4\sqrt{3}-4\sqrt{3}=0\sqrt{3}=0$
   c. $2\sqrt{10}+5\sqrt{10}=7\sqrt{10}$
   d. $3\sqrt{2}+4\sqrt{5}$ can't combine since $\sqrt{2}$ and $\sqrt{5}$ are not like square roots.
   e. $7\sqrt{2}+4\sqrt{5}+2\sqrt{2}-2\sqrt{5}=\left(7\sqrt{2}+2\sqrt{2}\right)+\left(4\sqrt{5}-2\sqrt{5}\right)=9\sqrt{2}+2\sqrt{5}$
   f. $5+2\sqrt{3}+4\sqrt{3}-2=(5-2)+\left(2\sqrt{3}+4\sqrt{3}\right)=3+6\sqrt{3}$

Sometimes square roots must be simplified before they can be added or subtracted.

EXAMPLE 1: $\sqrt{12}+\sqrt{27}$

SOLUTION:		
	$=\sqrt{4\cdot3}+\sqrt{9\cdot3}$	$12=4\cdot3$ and $27=9\cdot3$
	$=\sqrt{4}\sqrt{3}+\sqrt{9}\sqrt{3}$	Applying Rule 1
	$=2\sqrt{3}+3\sqrt{3}$	Evaluate $\sqrt{4}$ and $\sqrt{9}$
	$=5\sqrt{3}$	Add like square roots

EXAMPLE 2: $\sqrt{500}+\sqrt{180}$

SOLUTION:		
	$=\sqrt{5\cdot100}+\sqrt{5\cdot36}$	$500=5\cdot100$ and $180=5\cdot36$
	$=\sqrt{5}\sqrt{100}+\sqrt{5}\sqrt{36}$	Apply Rule 1
	$=\sqrt{5}\times10+\sqrt{5}\times10$	Evaluate $\sqrt{100}$ and $\sqrt{36}$
	$=10\sqrt{5}+6\sqrt{5}$	Rewrite so the coefficient is to left of square root
	$=16\sqrt{5}$	Add like square roots

153

EXAMPLE 3: $4\sqrt{24} + 5\sqrt{54}$

SOLUTION: $\quad = 4\sqrt{4 \cdot 6} + 5\sqrt{9 \cdot 6}$ $\qquad$ $24 = 4 \cdot 6$ and $54 = 9 \cdot 6$

$\qquad\qquad = 4\sqrt{4}\sqrt{6} + 5\sqrt{9}\sqrt{6}$ $\qquad$ Apply Rule 1

$\qquad\qquad = (4 \cdot 2)\sqrt{6} + (5 \cdot 3)\sqrt{6}$ $\qquad$ Evaluate $\sqrt{4}$ and $\sqrt{9}$

$\qquad\qquad = 8\sqrt{6} + 15\sqrt{6}$

$\qquad\qquad = 23\sqrt{6}$ $\qquad\qquad$ Add like square roots

EXAMPLE 4: $5\sqrt{48} - \sqrt{75}$

SOLUTION: $\quad = 5\sqrt{16 \cdot 3} - \sqrt{25 \cdot 3}$ $\qquad$ $48 = 16 \cdot 3$ and $75 = 25 \cdot 3$

$\qquad\qquad = 5\sqrt{16}\sqrt{3} - \sqrt{25}\sqrt{3}$ $\qquad$ Apply Rule 1

$\qquad\qquad = (5 \cdot 4)\sqrt{3} - 5\sqrt{3}$ $\qquad$ Evaluate $\sqrt{16}$ and $\sqrt{25}$

$\qquad\qquad = 20\sqrt{3} - 5\sqrt{3}$

$\qquad\qquad = 15\sqrt{3}$ $\qquad\qquad$ Subtract like square roots

EXAMPLE 5: Indicate the correct choice. $\sqrt{20} + \sqrt{45}$

$\qquad$ **(a)** $\sqrt{120}$ $\qquad$ **(b)** $\sqrt{125}$ $\qquad$ **(c)** 65 $\qquad$ **(d)** $\sqrt{65}$ $\qquad$ **(e)** not given

SOLUTION: $\sqrt{20} + \sqrt{45}$

$\qquad\qquad = \sqrt{4 \cdot 5} + \sqrt{9 \cdot 5}$ $\qquad$ $20 = 4 \cdot 5$ and $45 = 9 \cdot 5$

$\qquad\qquad = \sqrt{4}\sqrt{5} + \sqrt{9}\sqrt{5}$ $\qquad$ Apply Rule 1

$\qquad\qquad = 2\sqrt{5} + 3\sqrt{5}$ $\qquad$ Evaluate $\sqrt{4}$ and $\sqrt{9}$

$\qquad\qquad = 5\sqrt{5}$ $\qquad\qquad$ Add like square roots

$\qquad\qquad = \sqrt{25}\sqrt{5}$ $\qquad\qquad$ Since $5 = \sqrt{25}$

$\qquad\qquad = \sqrt{25 \cdot 5} = \sqrt{125}$ $\qquad$ Apply Rule 1 in reverse order.

$\qquad$ Therefore, choice b is correct.

Exercises: Adding and Subtracting Like Square Roots

For 1 – 9, evaluate.

1. $\sqrt{72}+\sqrt{32}$
    (a) $\sqrt{104}$ (b) 104   (c) $10\sqrt{2}$ (d) $10+\sqrt{2}$ (e) not given

2. $2\sqrt{45}+5\sqrt{20}$
    (a) $7\sqrt{65}$ (b) $10\sqrt{5}$ (c) $16\sqrt{5}$ (d) $6\sqrt{5}$ (e) not given

3. $\sqrt{8}+\sqrt{18}$
    (a) $\sqrt{50}$ (b) $\sqrt{26}$ (c) $10\sqrt{2}$ (d) $10+\sqrt{2}$ (e) not given

4. $3\sqrt{48}-2\sqrt{75}+3\sqrt{49}-2\sqrt{25}$
    (a) $21+22\sqrt{3}$ (b) $11+22\sqrt{3}$ (c) $12\sqrt{3}+21$
    (d) $\sqrt{3}+11$ (e) not given

5. $\sqrt{75}-\sqrt{20}$
    (a) $\sqrt{27}$ (b) $5\sqrt{3}$ (c) $3\sqrt{5}$ (d) $\sqrt{55}$ (e) not given

6. $\left(4\sqrt{7}\right)^{2}$
    (a) $16\sqrt{7}$ (b) 112   (c) 28   (d) 416   (e) not given

7. $16+4\sqrt{5}-3\sqrt{5}-6$
    (a) $10+\sqrt{5}$ (b) $10+7\sqrt{5}$ (c) $20\sqrt{5}-9$ (d) $10-2\sqrt{5}$ (e) not given

155

8. $\sqrt{75} + 3\sqrt{2}$

    (a) $5\sqrt{3} + 3\sqrt{2}$ **(b)** $8\sqrt{6}$ **(c)** $8\sqrt{5}$ **(d)** $\sqrt{78}$ **(e)** not given

9. $\dfrac{\sqrt{75}}{2} + \dfrac{5\sqrt{2}}{4}$

    (a) $\dfrac{10\sqrt{3} + 5\sqrt{2}}{4}$ **(b)** $\dfrac{15\sqrt{5}}{4}$ **(c)** $\dfrac{5}{2}\sqrt{3} + \dfrac{10}{4}$

    (d) $\dfrac{\sqrt{80}}{2} + \dfrac{\sqrt{2}}{4}$ **(e)** not given

10. If $\sqrt{a}$ is less than 9 but more than 8, which of the following is true?

    (a) a is less than 81   **(b)** a is less than 64

    (c) a is more than 81   **(d)** a is more than 164   **(e)** not given

11. If $\sqrt{n}$ is between 6 and 7, which of the following is true?

    (a) n is more than 50   **(b)** n is more than 25

    (c) n is less than 25   **(d)** n is less than 20   **(e)** not given

12. If $\sqrt{n}$ is between 6 and 7, which of the following is true?

    (a) n is between 0 and 5   **(b)** n is between 0 and 4

    (c) n is between 36 and 49   **(d)** n is between 6 and 7   **(e)** not given

In 13 – 20, evaluate.

13. $3\sqrt{50} + \sqrt{32}$

    (a) $\sqrt{482}$ **(b)** $6\sqrt{2} + 2$ **(c)** $19\sqrt{2}$ **(d)** $3\sqrt{82}$ **(e)** $9\sqrt{32}$

156

14. $\left(2\sqrt{3}\right)^2 + \sqrt{144}$

    (a) 24   **(b)** 12   **(c)** $6\sqrt{3}$ **(d)** $12\sqrt{8}$ **(e)** not given

15. $\dfrac{\sqrt{32}}{4} + \left(\sqrt{6}\right)^2$

    (a) $6+\sqrt{2}$ **(b)** $\dfrac{\sqrt{38}}{4}$ **(c)** $\sqrt{8}+6$ **(d)** $\dfrac{1}{4}\sqrt{32}$ **(e)** not given

16. $\left(3\sqrt{2}\right) - 2\sqrt{8}$

    (a) $-\sqrt{2}$ **(b)** $\sqrt{6}$ **(c)** $-3\sqrt{2}$ **(d)** $2\sqrt{2}$ **(e)** not given

17. $3\sqrt{8} + 6\sqrt{18} - 8\sqrt{32}$

    (a) $\sqrt{18}$ **(b)** $4\sqrt{2}$ **(c)** $-8\sqrt{2}$ **(d)** $5\sqrt{8}$ **(e)** not given

18. $3\sqrt{8} + 6\sqrt{8} - 10\sqrt{8}$

    (a) $-2\sqrt{8}$ **(b)** $-2\sqrt{2}$ **(c)** $4+3\sqrt{2}$ **(d)** $5\sqrt{2}$ **(e)** not given

19. $3\sqrt{8} + 6\sqrt{12}$

    (a) $9\sqrt{8}$ **(b)** $9\sqrt{12}$ **(c)** $9\sqrt{20}$ **(d)** $6\sqrt{2}+12\sqrt{3}$ **(e)** not given

20. $4\sqrt{8} - 2\sqrt{10}$

    (a) $2\sqrt{2}$ **(b)** $4\sqrt{2}$ **(c)** $8\sqrt{2}-4\sqrt{5}$ **(d)** $5\sqrt{5}$ **(e)** not given

Chapter 5

Ratios and Proportions

SECTION A. Ratios

A **ratio** is a comparison of two quantities that have the same units.

EXAMPLE 1:   Compare 12 quarts to 6 quarts as a ratio.

SOLUTION: $\dfrac{12\text{ quarts}}{6\text{ quarts}} = \dfrac{12}{6} = \dfrac{2}{1}$

or 12 quarts : 6 quarts
        12 : 6 = 2 : 1
or 12 quarts to 6 quarts
        12 to 6 = 2 to 1

The ratios of three or more quantities may be expressed as a **continued ratio**, which is a ratio statement combining two or more separate ratios.

EXAMPLE 2:   3 : 5 : 8

SOLUTION: Combines the ratios 3 : 5, 5: 8, and 3: 8.

Note:
    1. A ratio can be written in 3 different ways:
        a. as a fraction;
        b. as 2 numbers separated by a colon (:);
        c. as two number separated by the word "to".
    2. The units are not written.
    3. When written as a fraction, it should be reduced to its lowest terms.

If we are asked to find the ratio of quantities with different units, we must first express the quantities in the same units and then find the ratio.

EXAMPLE 3:   Find the ratio of 4 nickels to a quarter.

SOLUTION: We note that 4 nickels = 20¢. 1 quarter = 25¢, therefore the ratio is $\dfrac{20}{25} = \dfrac{4}{5}$ or 4 : 5,

or 4 to 5.

EXAMPLE 4: Find the ratios
   a. 5 inches to 2 feet
   b. 2 hours to 50 minutes
   c. 40 seconds to 3 minutes

SOLUTION:

   a. $\dfrac{5 \text{ in.}}{24 \text{ in.}} = \dfrac{5}{24}$

   b. $\dfrac{120 \text{ min.}}{50 \text{ min.}} = \dfrac{12}{5}$

   c. $\dfrac{40 \text{ sec.}}{180 \text{ sec.}} = \dfrac{2}{9}$

EXAMPLE 5: A company spends $2000 a month on office supplies and $4000 a month on electricity. Find the ratio of the monthly cost of office supplies to the monthly cost of electricity, and express this ratio in three different forms.

SOLUTION: $\dfrac{2000}{4000} = \dfrac{1}{2}$; 2000:4000 = 1:2; 2000 to 4000 = 1 to 2

EXAMPLE 6: The cost of building a deck was $250 for labor and $450 for materials. What is the ratio as a fraction of the cost of materials to the total cost for labor and materials?

SOLUTION: Total cost: $250 + 450 = \$700$; cost of materials: $450. Therefore, the desired ratio

is $\dfrac{450}{700} = \dfrac{9 \times \overset{1}{\cancel{50}}}{14 \times \underset{1}{\cancel{50}}} = \dfrac{9}{14}$.

EXAMPLE 7: 15 women and 21 men are enrolled in a history class. What is
   a. the ratio of men to women,
   b. the ratio of women to men?

SOLUTION:

   a. $\dfrac{\text{men}}{\text{women}} = \dfrac{21}{15} = \dfrac{7 \times \overset{1}{\cancel{3}}}{5 \times \underset{1}{\cancel{3}}} = \dfrac{7}{5}$

   b. $\dfrac{\text{women}}{\text{men}} = \dfrac{15}{21} = \dfrac{5 \times \overset{1}{\cancel{3}}}{7 \times \underset{1}{\cancel{3}}} = \dfrac{5}{7}$

159

Exercises: Ratios

For 1 – 9, find the ratio of each of the following and express it as a fraction in simplest form.

1. 10 to 15

   (a) $\frac{2}{3}$ (b) $\frac{10}{15}$ (c) $\frac{5}{10}$ (d) $\frac{1}{2}$ (e) $\frac{1}{2}$

2. 4 minutes to 36 seconds

   (a) $\frac{1}{9}$ (b) $\frac{4}{36}$ (c) $\frac{240}{36}$ (d) $\frac{20}{3}$ (e) $\frac{60}{9}$

3. 2 hours to 90 minutes

   (a) $\frac{120}{90}$ (b) $\frac{2}{90}$ (c) $\frac{1}{45}$ (d) $\frac{4}{3}$ (e) $\frac{9}{12}$

4. 120 inches to 1 foot

   (a) $\frac{120}{1}$ (b) $\frac{120}{12}$ (c) $\frac{10}{1}$ (d) $\frac{1}{120}$ (e) $\frac{1}{10}$

5. 4 months to 1 year

   (a) $\frac{4}{1}$ (b) $\frac{1}{3}$ (c) $\frac{1}{4}$ (d) $\frac{3}{1}$ (e) $\frac{120}{365}$

6. 4 ounces to 3 pounds

   (a) $\frac{4}{3}$ (b) $\frac{3}{4}$ (c) $\frac{1}{12}$ (d) $\frac{4}{48}$ (e) $\frac{3}{4}$

7. 60 centimeters to 2 meters

   (a) $\frac{60}{2}$ (b) $\frac{30}{1}$ (c) $\frac{3}{10}$ (d) $\frac{60}{200}$ (e) not given

8. 75 : 96

   (a) $\frac{75}{96}$ (b) $\frac{25}{3}$ (c) $\frac{50}{6}$ (d) $\frac{3}{25}$ (e) not given

160

9. 0.96 to 75

    (a) $\dfrac{96}{7500}$　　　**(b)** $\dfrac{96}{75}$　　　**(c)** $\dfrac{32}{2500}$　　　**(d)** $\dfrac{32}{25}$　　　**(e)** not given

For 10 – 11, the price of a \$10.50 shirt was increased by \$2.50.

10. What is the ratio of the new price to the old price?

    (a) $\dfrac{10.5}{2.5}$　　　**(b)** $\dfrac{2.50}{10.50}$　　　**(c)** $\dfrac{26}{21}$　　　**(d)** $\dfrac{21}{26}$　　　**(e)** not given

11. What is the ratio of the old price to the new price?

    (a) $\dfrac{21}{26}$　　　**(b)** $\dfrac{10.5}{2.5}$　　　**(c)** $\dfrac{2.50}{10.50}$　　　**(d)** $\dfrac{8}{13}$　　　**(e)** not given

For 12 – 13, write each ratio as a fraction and simplify. A mixture contains 30 milliliters of water and 20 milliliters of chlorine.

12. State the ratio of chlorine to water.

    (a) $\dfrac{20}{50}$　　　**(b)** $\dfrac{30}{20}$　　　**(c)** $\dfrac{2}{3}$　　　**(d)** $\dfrac{3}{2}$　　　**(e)** not given

13. State the ratio of water to chlorine.

    (a) $\dfrac{20}{50}$　　　**(b)** $\dfrac{30}{20}$　　　**(c)** $\dfrac{2}{3}$　　　**(d)** $\dfrac{3}{2}$　　　**(e)** not given

For 14 – 15, write each ratio as a fraction and simplify. The class consisted of 35 juniors and 20 freshmen.

14. State the ratio of juniors to freshmen.

    (a) $\dfrac{7}{4}$　　　**(b)** $\dfrac{35}{55}$　　　**(c)** $\dfrac{20}{55}$　　　**(d)** $\dfrac{20}{35}$　　　**(e)** $\dfrac{4}{7}$

15. State the ratio of freshmen to juniors.

    (a) $\dfrac{7}{4}$　　　**(b)** $\dfrac{35}{55}$　　　**(c)** $\dfrac{20}{55}$　　　**(d)** $\dfrac{20}{35}$　　　**(e)** $\dfrac{4}{7}$

For 16 – 17, write each ratio as a fraction and simplify. The school team had a record of 24 wins and 9 losses.

16. State the ratio of wins to losses.

(a) $\frac{24}{33}$ (b) $\frac{24}{13}$ (c) $\frac{33}{24}$ (d) $\frac{8}{3}$ (e) not given

17. State the ratio of losses to wins.

(a) $\frac{24}{33}$ (b) $\frac{24}{13}$ (c) $\frac{33}{24}$ (d) $\frac{8}{3}$ (e) not given

For 18 – 20, on a map, the cities A, B, C, and D are situated on a straight line such that the distance from city A to city B, from city B to city C, and from city C to city D are all the same. What is the ratio of the distance

18. from city B to city C to the distance from city A to city B?

(a) $\frac{1}{1}$ (b) $\frac{1}{2}$ (c) $\frac{1}{3}$ (d) $\frac{2}{3}$ (e) not given

19. from city A to city B to the distance from city A to city C?

(a) $\frac{1}{1}$ (b) $\frac{1}{2}$ (c) $\frac{1}{3}$ (d) $\frac{2}{3}$ (e) not given

20. from city A to city B to the distance from city A to city D?

(a) $\frac{1}{1}$ (b) $\frac{1}{2}$ (c) $\frac{1}{3}$ (d) $\frac{2}{3}$ (e) not given

SECTION B. Rate

A **rate** is a comparison of two quantities that have different units, and one unit cannot be converted into the other. Rates are written as a fraction. We include the units when we write a rate.

EXAMPLE 1: A car drove 20 miles in four hours. Find the distance to time rate.

SOLUTION: $\dfrac{20 \text{ miles}}{4 \text{ hours}} = \dfrac{5 \text{ miles}}{1 \text{ hour}}$ usually read as 5 miles for each hour or 5 miles per hour (mph).

Note:
1. the units are written as part of the rate,
2. the rate is in simplest form when the two numbers forming the rate have no common factors.

EXAMPLE 2: Find the rate.
  a. 6 feet to 4 seconds
  b. 50 milligrams to 25 seconds
  c. 3 bushels from 8 trees
  d. $16 for 4 items

SOLUTION:

a. $\dfrac{6 \text{ feet}}{4 \text{ seconds}} = \dfrac{3 \text{ feet}}{2 \text{ seconds}}$

b. $\dfrac{50 \text{ milligrams}}{25 \text{ seconds}} = \dfrac{2 \text{ milligrams}}{1 \text{ second}}$

c. $\dfrac{3 \text{ bushels}}{8 \text{ trees}}$

d. $\dfrac{\$16}{4 \text{ items}} = \dfrac{\$4}{1 \text{ items}}$

A **unit rate** is a rate in which the number in the denominator is 1. The average rate of speed is usually described as the distance traveled per unit time such as mile per hour, feet per second. Similarly the average cost per item is the unit cost, i.e. the cost for one item. Thus to find the average rate of speed, we divide the total distance by the total time spent traveling. To find the cost per item, we divide the total cost by the number of items.

163

EXAMPLE 3:   A car traveled 130 miles in 4 hours. What is its unit rate?

SOLUTION: $\dfrac{130 \text{ miles}}{4 \text{ hours}} = \dfrac{32.5 \text{ miles}}{1 \text{ hour}}$ or 32.5 miles per hour usually written as 32.5 mph.

EXAMPLE 4:   The cost of 3 cans of vegetables is \$1.50. What is the unit cost?

SOLUTION: $\dfrac{\$1.50}{3 \text{ cans}} = 50¢$ per can.

EXAMPLE 5:   Bob runs 2 miles in 8 minutes. How many miles per hour is he moving?

SOLUTION: He is moving $\dfrac{2 \text{ miles}}{8 \text{ minutes}} = \dfrac{1}{4}$ miles per minute, so that in an hour, or 60 minutes,

he is moving $\dfrac{1}{4} \times 60 = 15$ miles. Therefore, Bob is moving 15 mph.

EXAMPLE 6:   A fruit store sells 3 apples for 97¢. What is the cost per apple to the nearest tenth of a cent?

SOLUTION: Divide the cost of all 3 apples (97¢) by the number of apples (3).
$\dfrac{97¢}{3 \text{ apples}} = 32.3¢$. Therefore, the unit cost is 32.3¢ per apple.

EXAMPLE 7:   Which is a better buy: 4 cans for 84¢ or 5 cans for 70¢?

SOLUTION: 4 cans for 84¢: unit cost is $\dfrac{84}{4} = 21¢$ per can.

5 cans for 70¢: unit cost is $\dfrac{70}{5} = 14¢$ per can. Therefore, 5 cans for 70¢ is a better buy.

EXAMPLE 8:   Who is traveling at the higher rate of speed: person A traveling 80 miles in 5 hours, or person B traveling 96 miles in 6 hours?

SOLUTION: Person A: $\dfrac{80 \text{ miles}}{5 \text{ hours}} = 16$ mph. Person B: $\dfrac{96 \text{ miles}}{6 \text{ hours}} = 16$ mph. Therefore, person A and B are going at the same rate.

EXAMPLE 9:   A car needed 29.4 gallons of gas to travel 565.6 miles. How many miles per gallon did the car average. Round to the nearest tenth.

SOLUTION: Distance ÷ total number of gallons $= \dfrac{565.6 \text{ miles}}{29.4 \text{ gallons}} = 19.2$ miles per gallon.

EXAMPLE 10:   A grocer bought 100 boxes of paper plates for \$1500 and sold them for \$1800. What was his profit per box?

SOLUTION: Total profit for all 100 boxes ÷ number of boxes $= \dfrac{\$1800 - \$1500}{100 \text{ boxes}} = \dfrac{\$300}{100} = \$3$.
Therefore, the grocer's profit per box was \$3.

EXAMPLE 11:   If x peaches cost a total of b cents, then how many cents would c peaches cost?

SOLUTION: Divide the total cost by the number of peaches. $\dfrac{b}{x}$ is the cost of each peach.

Therefore, $c \cdot \dfrac{b}{x}$ is the cost of c peaches.

Exercises: Rate

For 1 – 2, what is the unit rate in calories per gram of fat?

1. 410 calories for 19 grams of fat.

2. 205 calories for 7 grams of fat.

For 3 – 4, what is the unit rate in miles per gallon?

3. Traveling 315 miles on 14 gallons of gas.

4. Traveling 405 miles on 18 gallons of gas.

For 5 – 6, what is the unit rate in dollars per hour?

5. Earning $304 in 38 hours.

6. Earning $455 in 35 hours.

For 7 – 8, what is the unit rate in miles per hour (mph)?

7. Traveling 320 miles in 6 hours.

8. Traveling 410 miles in 7 hours.

9. Bob drives 280.5 miles on 11.5 gallons of gas. Find the approximate number of miles per gallon (mpg) that Bob averaged.
    (a) 28 mpg   **(b)** 27 mpg   **(c)** 26 mpg   **(d)** 25 mpg   **(e)** 24 mpg

10. Sue drove 252 miles in 4.5 hours. Find the number of miles per hour (mph) Sue averaged.
    (a) 54 mph   **(b)** 55 mph   **(c)** 56 mph   **(d)** 58 mph   **(e)** 52 mph

11. A spaceship uses 35,500 gallons of fuel in 2.5 minutes. How much fuel does the spaceship use in 1 minute?
    (a) 14,000 gallons   **(b)** 14,100 gallons   **(c)** 14,200 gallons
    (d) 14,300 gallons   **(e)** 14,400 gallons

12. A teacher works for 5 months and makes \$12,500. What is the teacher's wage per month?
    (a) \$2300   **(b)** \$2400   **(c)** \$2500   **(d)** \$2600   **(e)** \$2700

13. A store bought 500 shirts for \$6500 and then sold all for a total of \$8500. What was the store's profit per shirt?
    (a) \$2000.00   **(b)** \$5.00   **(c)** \$4.00   **(d)** \$10.00   **(e)** \$6.50

14. The distance from the sun to Earth is 93 million miles. It takes 500 seconds for the light from the sun to reach Earth. Find the speed that light travels per second.
    (a) 186,000 mi/sec   **(b)** 180,000 mi/sec   **(c)** 1,800,000 mi/sec
    (d) 1800 mi/sec   **(e)** 18,600 mi/sec

15. A satellite travels 420,200 miles per day. How far does it travel in per hour? Round to the nearest mile.
    (a) 175,00 mi/hr   **(b)** 18,000 mi/hr   **(c)** 17,508 mi/hr
    (d) 17,600 mi/hr   **(e)** 17,625 mi/hr

16. Carol rides her bike 2 miles in 9 minutes. How many miles per hour (mph) is she going? Round to the nearest mile.
    (a) 12 mph   **(b)** 13 mph   **(c)** 14 mph   **(d)** 15 mph   **(e)** 30 mph

17. If a train is moving at a rate of 1 mile in 40 seconds, how many miles per hour (mph) is it traveling?
    (a) 80 mph   **(b)** 180 mph   **(c)** 90 mph   **(d)** 40 mph   **(e)** 120 mph

18. A plane flies 200 miles in 52 minutes. About how many miles per hour (mph) does it average? Round to the nearest mile.

(a) 200 mph   (b) 240 mph   (c) 230 mph   (d) 231 mph   (e) 225 mph

19. A car travels 616 miles on 28 gallons of gas. Find how many miles the car can drive using one gallon of gas.

(a) 26 miles   (b) 24 miles   (c) 22 miles   (d) 20 miles   (e) 18 miles

20. A pulley makes 63 complete rotations in 18 seconds. How many rotations per second does the pulley make?

(a) 3   (b) 3.5   (c) 3.8   (d) 4.0   (e) 5

21. Sue found a book club that charges $108 for 9 copies of a book. What is the cost per copy?

(a) $11   (b) $10   (c) $12   (d) $10   (e) $9

For 22 – 24, the *XY* Robertson Academy of Arts school has a staffing policy requiring that for every 90 students there are 5 instructors, and for every 30 students there are 2 tutors.

22. How many students per instructor does the academy have?

(a) 20   (b) 19   (c) 18   (d) 90   (e) 5

23. How many students per tutor does the academy have?

(a) 20   (b) 25   (c) 30   (d) 15   (e) 26

24. How many tutors are needed to satisfy the staffing policy if there are 90 students in the academy?

(a) 3   (b) 4   (c) 5   (d) 6   (e) 7

For 25 – 27, the 3-Star Insurance Group requires each office to have 4 insurance agents for every 620 clients and 5 clerical staff members for every 310 clients.

25. How many clients per agent does the group have?
    (a) 160   **(b)** 155   **(c)** 150   **(d)** 130   **(e)** 120

26. How many clients per clerical staff member does the group have?
    (a) 70   **(b)** 65   **(c)** 62   **(d)** 60   **(e)** 55

27. How many clerical staff members would be required for an office that has 930 clients?
    (a) 20   **(b)** 15   **(c)** 10   **(d)** 5   **(e)** 3

For 28 – 29, the Crystal Shop has their Gold Lac crystal wine glasses on sale. A box of 8 glasses is $96, and a box of 6 is $78.

28. Find each unit price.
    (a) box of 8 is $10 per glass; box of 6 is $14 per glass
    (b) box of 8 is $12 per glass; box of 6 is $13 per glass
    (c) box of 8 is $13 per glass; box of 6 is $12 per glass
    (d) box of 8 is $7.50 per glass; box of 6 is $5.50 per glass
    (e) not given

29. Which is the better buy?
    (a) Box of 8   **(b)** Box of 6   **(c)** Both are same   **(d)** not given

For 30 – 31, the tanning salon has a special on their tanning sessions: 12 session for $96 or 15 sessions for $135.

30. Find each unit price.
    (a) 12 sessions is $7 per session; 15 sessions is $8 per session
    (b) 12 sessions is $8 per session; 15 sessions is $7 per session
    (c) 12 sessions is $8 per session; 15 sessions is $9 per session
    (d) 12 sessions is $9 per session; 15 sessions is $8 per session
    (e) not given

31. Which is the better deal?
    (a) 12 sessions   **(b)** 15 sessions   **(c)** Both are same   **(d)** not given

SECTION C. Proportions

A **proportion** is the equality of two ratios or rates.

For example,

two numerators

1. $\dfrac{\$30}{3 \text{ hours}} = \dfrac{\$90}{9 \text{ hours}}$ 4 numbers are involved in every proportion

two denominators

same units

2. $\dfrac{30 \text{ miles}}{5 \text{ gallons}} = \dfrac{60 \text{ miles}}{10 \text{ gallons}}$

same units

3. $\dfrac{6 \text{ feet}}{3 \text{ seconds}} = \dfrac{12 \text{ feet}}{6 \text{ seconds}}$

Note:
1. The units, if any, of the numerator are the same and the units, if any, of the denominator are the same.
2. A proportion is true when the two ratios are equal.
3. A proportion is true when one ratio can be obtained from the other by multiplying both numerator and denominator by the same non-zero number.
4. A proportion is true if the cross products are equal. By **cross products**, we mean the denominator of one fraction times the numerator of the other fraction.

   For example, $\dfrac{6}{3} = \dfrac{12}{6}$ is a true proportion because

   a. the ratio $\dfrac{6}{3} = 2$ and the ratio $\dfrac{12}{6} = 2$. Thus both ratios $\dfrac{6}{3}$ and $\dfrac{12}{6}$ are equal.

   b. $\dfrac{6 \times 2}{3 \times 2} = \dfrac{12}{6}$, $\dfrac{12}{6}$ is obtained by multiplying numerator and denominator of $\dfrac{6}{3}$ by 2.

   c. $\dfrac{6}{3} \times \dfrac{12}{6} \begin{matrix} \rightarrow 3 \times 12 = 36 \\ \rightarrow 6 \times 6 = 36 \end{matrix} \Big\}$ the cross products are equal.

5. A proportion is not true if the cross products are not equal.

   For example, $\dfrac{6}{4} = \dfrac{12}{5}$ is not a true proportion because

   $\dfrac{6}{4} \times \dfrac{12}{5} \begin{matrix} \rightarrow 4 \times 12 = 48 \\ \rightarrow 6 \times 5 = 30 \end{matrix} \Big\}$ the cross products are not equal.

170

Exercises: Proportions

For 1 – 20, determine whether each of the following is a true proportion.

1. $\dfrac{4}{8} = \dfrac{2}{4}$

2. $\dfrac{13}{26} = \dfrac{19}{38}$

3. $\dfrac{1}{4} = \dfrac{5}{20}$

4. $\dfrac{7}{11} = \dfrac{8}{12}$

5. $\dfrac{72}{4} = \dfrac{54}{3}$

6. $\dfrac{9}{6} = \dfrac{7}{5}$

7. $\dfrac{6 \text{ minutes}}{5 \text{ miles}} = \dfrac{12 \text{ minutes}}{9 \text{ miles}}$

8. $\dfrac{\$65}{3 \text{ days}} = \dfrac{\$25}{2 \text{ days}}$

9. $\dfrac{80 \text{ miles}}{2 \text{ hours}} = \dfrac{120 \text{ miles}}{3 \text{ hours}}$

10. $\dfrac{4}{9} = \dfrac{28}{63}$

11. $\dfrac{63}{9} = \dfrac{28}{4}$

12. $\dfrac{63}{28} = \dfrac{9}{4}$

13. $\dfrac{3}{8} = \dfrac{18}{48}$

14. $\dfrac{48}{8} = \dfrac{18}{3}$

15. $\dfrac{48}{18} = \dfrac{3}{8}$

16. $\dfrac{72}{54} = \dfrac{4}{3}$

17. $\dfrac{3}{54} = \dfrac{4}{72}$

18. $\dfrac{9}{7} = \dfrac{6}{5}$

19. $\dfrac{1}{5} = \dfrac{4}{20}$

20. $\dfrac{38}{19} = \dfrac{26}{13}$

171

SECTION D. Writing and Reading a Proportion

When we write a proportion we must be sure that the units are in the appropriate position. One way to write a proportion is for the numerators to have the same units and the denominators to have the same units. In other words we write each fraction as we would a rate or a ratio. In the proportion $\frac{a}{b}=\frac{c}{d}$, a and c should have the same units, and b and d should also have the same units.

EXAMPLE 1:   Write the proportion to express the following: If 6 pounds of flour cost \$2, then 18 pounds will cost \$6.

SOLUTION:

$$\frac{6\text{ pounds}}{2\text{ dollars}}=\frac{18\text{ pounds}}{6\text{ dollars}}$$ We write pounds in the numerator.
We write dollars in the denominator.

EXAMPLE 2:   Write the proportion to express the following: If it takes 4 hours to drive 144 miles, it will take 6 hours to drive 216 miles.

SOLUTION:

$$\frac{4\text{ hours}}{144\text{ miles}}=\frac{6\text{ hours}}{216\text{ miles}}$$ We write hours in the numerator.
We write miles in the denominator.

The proportion $\frac{a}{b}=\frac{c}{d}$ is read "a is to b as c is to d."

For example, $\frac{4}{8}=\frac{1}{2}$ is read "4 is to 8 as 1 is to 2." And the proportion 3 is to 8 as 15 is to 40 is written as $\frac{3}{8}=\frac{15}{40}$.

172

Exercises: Writing and Reading a Proportion

For 1 – 20, write the proportion.

1. 4 is to 9 as 28 is to 63.

2. 16 is to 5 as 48 is to 15.

3. $\frac{1}{3}$ is to $\frac{1}{8}$ as $\frac{1}{4}$ is to $\frac{3}{32}$.

4. 6 is to 11 as 30 is to 55.

5. 3 is to 8 as 18 is to 48.

6. $\frac{1}{7}$ is to $\frac{1}{9}$ as $\frac{1}{6}$ is to $\frac{7}{54}$.

7. 12 is to 7 as 48 is to 28.

8. 2 is to 9 as 8 is to 36.

9. 9 is to 11 as 8 is to 12.

10. 3 pounds is to 1 dollar as 18 pounds is to 6 dollars.

11. If 2 printers are needed for 4 secretaries, 12 printers are needed for 24 administrative assistants.

12. If 2 cups of cereal contain 50 grams of carbohydrates, then 6 cups of cereal contain 150 grams of carbohydrates.

13. If 6 American dollars have a value of 4 British pounds, then 264 American dollars have a value of 176 British pounds.

14. If a pulley can complete $3\frac{1}{2}$ rotations in 2 minutes, it should complete 14 rotations in 8 minutes.

15. If Matt averages 4 baskets out of 7 free throws attempted in a basketball game, he should make 12 out of 21 free throws.

16. If Sal averages 2 baskets out of 5 free throws attempted in a basketball game, he should make 12 out of 30 free throws.

17. If 6 benches are needed to seat 24 people, 48 benches are needed to seat 192 people.

18. If 3 inches on a map represent 270 miles, 6 inches represent 540 miles.

19. If Jaime can drive 176 miles in his Toyota truck on 8 gallons of gas, then he should be able to drive 528 miles on 24 gallons of gas.

20. If it takes $2\frac{1}{4}$ yards of material to make 1 skirt, it will take 9 yards to make 4 skirts.

E. Solving $ax = b$

When we do not know the value of a number, we use a letter such as n, x, y, and z to represent that number. A letter that represents a number is called a **variable**. Variables, numbers, and combinations of variables and numbers such as x, $\sqrt{2}$, $x + 52$, $y + 52$, and $3 - x$ are called algebraic expressions, variable expressions or simply expressions.

In mathematics, there are several ways of indicating multiplication. We write the multiplication of 5 times 6 in the following ways:
$$(5)6, \quad 5(6), \quad 5 \cdot 6, \quad 5 \times 6, \quad (5)(6), \quad 5 * 6.$$
If two variables a and b are multiplied, we indicate this by writing ab with symbols between the a and b. If a number is multiplied by a variable, we write the number first with no symbols between the number and the variable. Thus $7x$ means "seven times a number." The numbers we multiply are called **factors**. The result of the multiplication is called the **product**.

For example,
1. $(7)5 = 35$ 7 and 5 are factors and 35 is the product.
2. $4x = 20$ 4 and x are factors and 20 is the product.

To evaluate variable expressions, we simply replace the variable in the expression with the given value and simplify.

EXAMPLE 1:   Evaluate $8x$ for $x = 4$ and $x = 5$.

SOLUTION:
$8(4) = 32$, thus when $x = 4$, $8x$ is equal to 32.
$8(5) = 40$, thus when $x = 5$, $8x$ is equal to 40.

Two expressions separated by an equal sign are called an **equation**. For example, $3x = 12$. To **solve** an equation, we must find a value, which is called the **solution**, for the variable in the equation that makes the equation a true statement. We can think of an equation as a question and the solution is the answer to this question.

For example, the equation $5x = 20$ can be interpreted as the question "fives times what number is equal to twenty?" The answer is 4, because $5 \cdot 4 = 20$. The number 4 is called the solution to the equation $5x = 20$ and is written $x = 4$. "The value of x is 4." The solution must make the equation a true statement, i.e. when the solution is substituted into the equation, the left and right-hand side of the equation must be equal.

175

EXAMPLE 2:   Is 5 a solution to the equation $6x = 30$?

SOLUTION:

$6x = 30$ Six times what number equals 30?

$\overset{?}{6(5) = 30}$ Evaluate $6x$ for $x = 5$.

$\overset{\sqrt{}}{30 = 30}$ We get a true statement.

We say that $x = 5$ is a solution to the equation $6x = 30$.

EXAMPLE 3:   Is 4 a solution to the equation $7x = 35$?

SOLUTION:

$7x = 35$ Seven times what number equals 35?

$\overset{?}{7(4) = 35}$ Replace the variable with 4 and simplify.

$\overset{?}{28 = 35}$ This is a false statement.

Since $28 = 35$ is not a true statement, so 4 is **not** a solution to $7x = 35$.

We can easily solve the equation of the form $ax = b$, i.e. a number, call it a, times a variable, call it x (or any other letter) = some number, call it b, by using the division principle.

The Division principle: If both sides of an equation are divided by the same nonzero number, the results on both sides are equal in value. Using the division principle, we divide both sides of the equation by the number multiplying the variable.

EXAMPLE 4:   Solve $6x = 30$.

SOLUTION:

$$\frac{\overset{1}{\cancel{6}}x}{\underset{1}{\cancel{6}}} = \frac{30}{6} \quad \text{Divide both sides by 6.}$$

$x = 5 \qquad \dfrac{6x}{6} = x \text{ and } \dfrac{30}{6} = 5$

To check our solution we must replace x by 5 and see whether we get a true statement.

$\overset{?}{6(5) = 30}$

$\overset{\sqrt{}}{30 = 30}$ True statement. Hence, $x = 5$ is indeed our solution.

EXAMPLE 5:   Solve $4x = 9$.

SOLUTION:

$$\frac{4x}{4} = \frac{9}{4}$$ Divide both sides by 4.

$$x = 2\frac{1}{4} \qquad \frac{\cancel{4}x}{\cancel{4}} = x \text{ and } \frac{9}{4} = 2\frac{1}{4}$$

To check our solution we must replace x by $2\frac{1}{4}$ and see whether we get a true statement.

$$4\left(2\frac{1}{4}\right) \overset{?}{=} 9$$

$$\overset{\surd}{9 = 9} \quad \text{True statement. Hence, } x = 2\frac{1}{4} \text{ is indeed our solution.}$$

EXAMPLE 6:   Solve $-3x = 30$.

SOLUTION:

$$\frac{-3x}{-3} = \frac{30}{-3}$$ Divide both sides by -3.

$$x = -10 \qquad \frac{\cancel{-3}x}{\cancel{-3}} = x \text{ and } \frac{30}{-3} = -10$$

To check our solution we must replace x by -10 and see whether we get a true statement.

$$-3(-10) \overset{?}{=} 30$$

$$\overset{\surd}{30 = 30} \quad \text{True statement. Hence, } x = -10 \text{ is indeed our solution.}$$

EXAMPLE 7:   Solve $7x = 42$.

SOLUTION:

$$\frac{7y}{7} = \frac{42}{7}$$ Divide both sides by 7.

$$y = 6 \qquad \frac{\cancel{7}y}{\cancel{7}} = y \text{ and } \frac{42}{7} = 6$$

To check our solution we must replace y by 6 and see whether we get a true statement.

$$7(6) \overset{?}{=} 42$$

$$\overset{\surd}{42 = 42} \quad \text{True statement. Hence, } y = 6 \text{ is indeed our solution.}$$

EXAMPLE 8: Solve $81 = 3x$.

SOLUTION:

$\dfrac{81}{3} = \dfrac{3x}{3}$ Divide both sides by 3.

$27 = x$ $\dfrac{81}{3} = 27$ and $\dfrac{3x}{3} = x$

To check our solution we must replace x by 27 and see whether we get a true statement.

$81 \overset{?}{=} 3(27)$

$81 = 81$ True statement. Hence, $x = 27$ is indeed our solution.

Exercises: Solving *ax* = *b*

For 1 – 24, solve and check.

1. $3x = 39$ **2.** $7y = -21$

3. $-3y = 15$ **4.** $42 = 6x$

5. $8x = 104$ **6.** $-19x = -76$

7. $5x = 65$ **8.** $2a = -32$

9. $-4x = 12$ **10.** $12 = 2y$

11. $9y = 135$ **12.** $-22y = -132$

13. $11x = 55$ **14.** $5m = -35$

15. $-7a = 49$ **16.** $18 = 9x$

17. $-15y = 165$ **18.** $55 = 5a$

19. $9x = 99$ **20.** $6y = -42$

21. $-5y = 20$ **22.** $56 = 14y$

23. $-13x = 156$ **24.** $44 = 4y$

179

SECTION F. Solving a Proportion

Sometimes one of the four number in a proportion is unknown. To solve the proportion, we find a number to replace the unknown so that the proportion is true.

EXAMPLE 1: $\frac{2}{7} = \frac{8}{?}$ Find the unknown value (which is denoted by ?) so that the proportion is true.

SOLUTION: Since we know that if the proportion is to be true, the cross products must be equal, i.e. the product of 2 and the unknown must be equal to $7 \times 8 = 56$. So the question is, what number times 2 is 56? After a little thought we realized that number must be 28.

We can replace the "?" in the above example with x and have an equation $\frac{2}{7} = \frac{8}{x}$. Cross multiplication yields $2x = 7(8) = 56$. To find x we use the procedure of the last section: divide both $2x$ and 56 by the number multiplying x, namely 2. Thus to solve for x:

$$\frac{2x}{2} = \frac{56}{2}$$
$$x = 28$$

EXAMPLE 2: Solve $\frac{24}{x} = \frac{12}{5}$

SOLUTION:

$$12x = 24(5) \quad \text{Cross multiply}$$
$$\frac{12x}{12} = \frac{(24)5}{12} \quad \text{Solve for } x$$
$$x = 10$$

EXAMPLE 3: What value of x solves the proportion $\frac{3}{2} = \frac{x}{8}$?

SOLUTION:

$$2x = 24 \quad \text{Cross products are equal}$$
$$\frac{2x}{2} = \frac{24}{2} \quad \text{Solve for } x$$
$$x = 12$$

We note that solving the proportion $\frac{3}{2} = \frac{x}{8}$ is equivalent to solving the equation $2x = 24$.

EXAMPLE 4: Solve and check $\dfrac{x}{21} = \dfrac{2}{7}$

SOLUTION:

$7x = 21(2)$ Form an equation by equating the cross products

$\dfrac{7x}{7} = \dfrac{42}{7}$ Divide both sides by 7

$x = 6$ $\dfrac{7x}{7} = x$ and $\dfrac{42}{7} = 6$

Check:

$\dfrac{6}{21} \overset{?}{=} \dfrac{2}{7}$ Replace x in the original problem by 6

$7 \times 6 \overset{?}{=} 21 \times 2$ Cross multiply to see whether the proportion is true.

$42 = 42$

EXAMPLE 5: Find the value of n and check $\dfrac{n}{18} = \dfrac{28}{72}$?

SOLUTION:

$72n = 28 \cdot 18$

$\dfrac{72n}{72} = \dfrac{28 \cdot 18}{72}$

$n = \dfrac{\cancel{4} \cdot 7 \cdot \cancel{18}}{\cancel{4} \cdot \cancel{18}}$

$n = 7$

Check:

$\dfrac{7}{18} \overset{?}{=} \dfrac{28}{72}$

$7 \cdot 72 \overset{?}{=} 18 \cdot 28$

$504 \overset{\checkmark}{=} 504$

Exercises: Solving a Proportion

For 1 – 5, solve for *x*.

1. $\dfrac{x}{6} = \dfrac{4}{8}$

2. $\dfrac{12}{x} = \dfrac{18}{9}$

3. $\dfrac{24}{12} = \dfrac{x}{10}$

4. $\dfrac{15}{10} = \dfrac{3}{x}$

5. $\dfrac{4 \text{ ounces}}{50 \text{ pounds}} = \dfrac{x \text{ ounces}}{100 \text{ pounds}}$

For 6 – 11, find the value of *n*.

6. $\dfrac{80 \text{ gallons}}{24 \text{ acres}} = \dfrac{20 \text{ gallons}}{n \text{ acres}}$

7. $\dfrac{n \text{ grams}}{15 \text{ liters}} = \dfrac{16 \text{ grams}}{45 \text{ liters}}$

8. $\dfrac{6 \text{ hours}}{150 \text{ miles}} = \dfrac{1 \text{ hour}}{n \text{ miles}}$

9. $\dfrac{12 \text{ women}}{18 \text{ men}} = \dfrac{n \text{ women}}{21 \text{ men}}$

10. $\dfrac{5 \text{ miles}}{4 \text{ kilometers}} = \dfrac{n \text{ miles}}{8 \text{ kilometers}}$

11. $\dfrac{25 \text{ miles}}{10 \text{ hours}} = \dfrac{20 \text{ miles}}{n \text{ hours}}$

12. If the ratio of 12 to x is 18 to 9, then what is x?
　　(a)　5　　　　　(b)　4　　　　　(c)　6　　　　　(d)　8　　　　　(e)　10

13. If the ratio of 15 to 10 is 3 to y, then what is y?
　　(a)　2　　　　　(b)　3　　　　　(c)　4　　　　　(d)　5　　　　　(e)　6

14. What is the equivalent equation that has to be solved in order to solve the proportion $\frac{6}{8} = \frac{24}{x}$?
　　(a)　$6 + x = 8 + 24$　　　　　(b)　$6x = 8(24)$　　　　　(c)　$6 + 8 = 24 + x$
　　(d)　$8x = 6(24)$　　　　　(e)　not given

15. What is the equivalent equation that has to be solved in order to solve the proportion $\frac{x}{5} = \frac{6000}{3}$?
　　(a)　$6000x = 15$　　　　　(b)　$3x = 5(6000)$　　　　　(c)　$5x = 3(6000)$
　　(d)　$x + 5 = 6003$　　　　　(e)　$x + 3 = 6000 + 5$

16. If $\frac{x}{30} = \frac{65}{120}$, then $x = ?$
　　(a)　$\frac{65}{120}$　　　　　(b)　$\frac{30}{9}$　　　　　(c)　$\frac{13}{8}$　　　　　(d)　$\frac{26}{8}$　　　　　(e)　$\frac{14}{9}$

17. What value of x solve the proportion $\frac{21}{x} = \frac{7}{15}$?
　　(a)　3　　　　　(b)　45　　　　　(c)　315　　　　　(d)　21　　　　　(e)　105

18. Two quantities A and B are always in the same ratio. If $A = 3$ when $B = 2$, what is the value of A when $B = 4$?
　　(a)　8　　　　　(b)　6　　　　　(c)　14　　　　　(d)　2　　　　　(e)　not given

19. What is the equivalent equation that has to be solved in order to solve the proportion $\frac{6}{x} = \frac{8}{36}$?
　　(a)　$6x = 8(36)$　　　　　(b)　$8x = 6(36)$　　　　　(c)　$6x = 36(8)$
　　(d)　$36x = 6(8)$　　　　　(e)　$36x = 8(6)$

20. If $\frac{1.7}{3.4} = \frac{x}{12}$, then $x = ?$
　　(a)　6　　　　　(b)　8.2　　　　　(c)　4.4　　　　　(d)　9　　　　　(e)　4

SECTION G. Solving Applied Problems Involving Proportions

When a situation involves a ratio or rate, we can use a proportion to find the solution. Let us examine a variety of applied problems that can be solved with proportions.

EXAMPLE 1: A large automobile dealership has found that for every 14 cars sold, 2 are brought back for major repairs. If the dealership sells 112 cars this month, approximately how many cars will be brought back for major repairs?

SOLUTION: We set up the proportion and then solve for the missing number in the proportion. We let n represent the total number of major repairs.

$$\frac{14 \text{ cars sold}}{2 \text{ need major repairs}} = \frac{112 \text{ total cars sold}}{n \text{ total major repairs}}$$

$$\frac{14}{2} = \frac{112}{n}$$

$14n = 2 \times 112$ Form a cross product.

$14n = 224$ Simplify.

$$\frac{14n}{14} = \frac{224}{14}$$ Divide both sides by 14.

$$n = 16$$

Approximately 16 cars will be brought back for major repairs.

EXAMPLE 2: Mary Lou's Catering has a policy that when planning a buffet there should be 18 desserts for every 15 people who will be attending the buffet. How many desserts should the catering company plan to serve at a buffet if 180 people are expected to attend?

SOLUTION:

$$\frac{18 \text{ desserts}}{15 \text{ people}} = \frac{n \text{ desserts}}{180 \text{ people}}$$

$$\frac{18}{15} = \frac{n}{180}$$

$18 \cdot 180 = 15 \cdot n$ Form a cross product.

$$\frac{18 \cdot 80}{15} = \frac{15n}{15}$$ Divide both sides by 15.

$$216 = n$$

216 desserts should be served.

EXAMPLE 3: Estelle has a fence in her yard around her vegetable garden, which is 6 feet wide and 8 feet long. The fence's dimensions are proportional to the garden's. What is the length of the fence if the width is 24 feet?

SOLUTION: First, we set up the proportion, letting the letter x represent the length of the fence.

$$\frac{\text{width of garden}}{\text{length of garden}} = \frac{\text{width of fence}}{\text{length of fence}}$$

$$\frac{6 \text{ ft}}{8 \text{ ft}} = \frac{24 \text{ ft}}{x \text{ ft}}$$

Now we solve for x.

$$\frac{6}{8} = \frac{24}{x}$$

$$6x = 8 \times 24$$

$$6x = 192$$

$$\frac{6x}{6} = \frac{192}{6}$$

$$x = 32 \qquad \text{The length of the fence is 32 feet.}$$

EXAMPLE 4: Refer to Example 3 to answer the following. If the width of the fence is 18 feet, what must the length be for the dimensions of the fence to be proportional to those of the garden?

SOLUTION: First, we set up the proportion, letting the letter x represent the length of the fence.

$$\frac{\text{width of garden}}{\text{length of garden}} = \frac{\text{width of fence}}{\text{length of fence}}$$

$$\frac{6 \text{ ft}}{8 \text{ ft}} = \frac{18 \text{ ft}}{x \text{ ft}}$$

Now we solve for x.

$$\frac{6}{8} = \frac{18}{x}$$

$$6x = 8 \cdot 18$$

$$\frac{6x}{6} = \frac{8 \cdot 18}{6}$$

$$x = 24 \qquad \text{The length of the fence is 24 feet.}$$

Since a proportion is the equality of ratios and a ratio is a fraction, we can reduce a ratio in a proportion without changing the value of the ratio. Therefore, we can reduce a proportion before we solve for the missing number in the proportion. If you can see that the ratio without the variable can be reduced, you may reduce it, and the answer will still be correct. Let's look at the proportion in Example 3, and observe what happens when we reduce the ratio $\frac{6}{8}$.

$$\frac{6}{8} = \frac{24}{x}$$

$$\frac{3}{4} = \frac{24}{x} \qquad \text{Reduce } \frac{6}{8} \text{ to } \frac{3}{4}$$

$$3x = 4 \times 24 \qquad \text{Cross multiply}$$

$$3x = 96$$

$$\frac{3x}{3} = \frac{96}{3}$$

$$x = 32 \qquad \text{We see that the answer is the same.}$$

EXAMPLE 5:   Two partners, Cleo and Julie, invest money in their small business at the ratio of 3 to 5, with Cleo investing the smaller amount. If Cleo invested $6000, how much did Julie invest?

SOLUTION: The ratio *3 to 5* represents Cleo's investment *to* Julie's investment.

$$\frac{3}{5} = \frac{\text{Cleo's investment of \$6000}}{\text{Julie's investment of \$}x}$$

$$\frac{3}{5} = \frac{6000}{x}$$

$$3x = 30,000$$

$$\frac{3x}{3} = \frac{30,000}{3}$$

$$x = 10,000$$

Julie invested $10,000 in their business.

EXAMPLE 6:   Refer to Example 5 to answer the following. Cleo and Julie also split the profits from the partnership in the same ratio, 3 to 5. If Cleo receives $2400 for her share of the profit, how much does Julie receive in profits?

SOLUTION: The ratio *3 to 5* represents Cleo's profits *to* Julie's profits.

$$\frac{3}{5} = \frac{\text{Cleo's profits of \$2400}}{\text{Julie's profits of \$}x}$$

$$\frac{3}{5} = \frac{2400}{x}$$

$$3x = 12,000$$

$$\frac{3x}{3} = \frac{12,000}{3}$$

$$x = 4000 \qquad \text{Julie's profit would be \$4000.}$$

EXAMPLE 7: A recipe for tuna casserole to serve 9 people calls for 4 pounds of tuna. Dave wants to make the casserole for 15 people. Using the proportion of that recipe, how many pounds of tuna does he need?

SOLUTION:

$$\frac{9 \text{ people}}{4 \text{ lbs tuna}} = \frac{15 \text{ people}}{x \text{ lbs of tuna}}$$

$$9x = 4 \cdot 15$$

$$\frac{9x}{9} = \frac{60}{9}$$

$$x = 6\frac{2}{3} \qquad \qquad 6\frac{2}{3} \text{ pounds of tuna are needed.}$$

EXAMPLE 8: A recipe requires 3 ingredients A, B, C in the volume ratio 2 : 3 : 4. If 6 pints of ingredient B are required, how many pints of ingredients A and C are required?

SOLUTION: For ingredient A:

$$\frac{2}{3} = \frac{x \text{ pints of ingredient } A}{6 \text{ pints of ingredient } B}$$

$$\frac{2}{3} = \frac{x}{6}$$

$$3x = 12$$

$$\frac{3x}{3} = \frac{12}{3}$$

$$x = 4 \qquad \qquad 4 \text{ pints of ingredient } A \text{ are required.}$$

For ingredient C:

$$\frac{3}{4} = \frac{6 \text{ pints of ingredient } B}{x \text{ pints of ingredient } C}$$

$$\frac{3}{4} = \frac{6}{x}$$

$$3x = 24$$

$$\frac{3x}{3} = \frac{24}{3}$$

$$x = 8 \qquad \qquad 8 \text{ pints of ingredient } C \text{ are required.}$$

EXAMPLE 9:   On a certain map, 1 inch represents 75 miles. If the distance between two cities is 200 miles, how many inches on the map should represent the distance between the two cities?

SOLUTION:

$$\frac{1 \text{ inch}}{75 \text{ miles}} = \frac{x \text{ inches}}{200 \text{ miles}}$$

$$\frac{1}{75} = \frac{x}{200}$$

$$75x = 200$$

$$\frac{75x}{75} = \frac{200}{75}$$

$$x = 2\frac{2}{3}$$

$2\frac{2}{3}$ inches should be the distance between these two cities on the map.

EXAMPLE 10:   Water consists, by mass, of 8 parts of oxygen and 1 part of hydrogen. How many grams of oxygen are there in 108 grams of water?

SOLUTION: Water consists of oxygen and hydrogen in ratio of 8 to 1. This means that for every 9 parts of water, 8 parts will by oxygen and 1 part is hydrogen so that $\dfrac{\text{\# of parts oxygen}}{\text{total parts of water}} = \dfrac{8}{9}$.

Therefore, we solve the proportion $\dfrac{x}{108 \text{ g}} = \dfrac{8}{9}$; or $9x = 8(108)$. Then $x = \dfrac{8(108)}{9} = 8 \cdot 12 = 96 \text{ g}$. In other words, in 108 grams of water, there are 96 grams of oxygen.

EXAMPLE 11:   John brings home the same paycheck each month. He spends $\dfrac{3}{5}$ of it and saves the rest. If John spends $180 of his paycheck each month, what is his monthly pay?

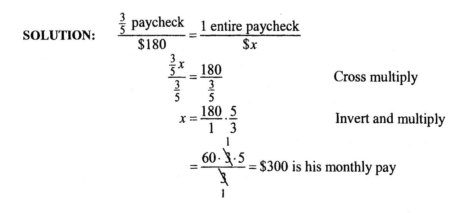

SOLUTION:

$$\frac{\frac{3}{5} \text{ paycheck}}{\$180} = \frac{1 \text{ entire paycheck}}{\$x}$$

$$\frac{\frac{3}{5}x}{\frac{3}{5}} = \frac{180}{\frac{3}{5}} \qquad \text{Cross multiply}$$

$$x = \frac{180}{1} \cdot \frac{5}{3} \qquad \text{Invert and multiply}$$

$$= \frac{60 \cdot \cancel{3} \cdot 5}{\cancel{3}} = \$300 \text{ is his monthly pay}$$

Exercises: Solving Applied Problems Involving Proportions

1. If 2 cups of cereal contain 50 grams of carbohydrates, how many grams of carbohydrates do 5 cups of cereal contain?
    (a) 50 g   (b) 100 g   (c) 125 g   (d) 250 g   (e) 75 g

2. If shirts are on sale at 3 for $25, how much will it cost to buy 12 shirts?
    (a) $75   (b) $100   (c) $50   (d) $110   (e) $125

3. If it takes Mason 25 minutes to water 2 rows of plants in the field, how long will it take him to water a field of 12 rows?
    (a) 75 min   (b) 100 min   (c) 150 min   (d) 175 min   (e) 110 min

4. If a 200-pound man can have 1000 milligrams of medicine a day, how much can a 120-pound woman have?
    (a) 200 mg   (b) 400 mg   (c) 600 mg   (d) 700 mg   (e) 450 mg

5. A school has a student to teacher ratio of 35 to 2. If the school has 875 students, how many teachers does it have?
    (a) 100   (b) 125   (c) 200   (d) 225   (e) 50

6. A baseball player gets 20 hits out of 50 times at bat. How many hits must she get in her next 150 times at bat to keep her batting average the same?
    (a) 50   (b) 60   (c) 70   (d) 80   (e) 85

7. In a scale drawing, a 210-foot-tall building is drawn 3 inches high. If another building is drawn 5 inches high, how tall is that building?
    (a) 300 ft.   (b) 325 ft.   (c) 350 ft.   (d) 400 ft.   (e) 420 ft.

8. If 100 grams of ice cream contain 15 grams of fat, how much fat is in 260 grams of ice cream?
    (a) 20 g   (b) 25 g   (c) 30 g   (d) 35 g   (e) 39 g

9. If 5 shares of a certain stock cost $160, how much will 12 shares cost?
    (a) $380   (b) $381   (c) $382   (d) $384   (e) $385

10. A recipe for a punch calls for 5 cups of water for every 2 cups of punch concentrate. How much water is needed with 8 cups of concentrate?
    (a) 10 cups   (b) 15 cups   (c) 20 cups   (d) 25 cups   (e) 30 cups

11. In a stock split, each person received 8 shares for each 5 shares that he or she held. If a person had 850 shares of stock in the company, how many shares did she receive in the stock split?
    (a) 1000   **(b)** 1200   **(c)** 1300   **(d)** 1350   **(e)** 1360

12. A 100-watt stereo system needs copper speaker wire that is 30 millimeters thick to handle the output of sound clearly. How thick would the speaker wire need to be if you had a 140-watt stereo and you wanted the same ratio of watts to millimeters?
    (a) 22 mm   **(b)** 32 mm   **(c)** 42 mm   **(d)** 52 mm   **(e)** 152 mm

13. Julio wants to put a fence around his rectangular pool, which is 12 feet wide and 18 feet long. If the size of the yard will only allow for a fence that is 30 feet long and Julio wants the dimensions of the fence to be proportional to those of the pool, how wide should the fence be?
    (a) 10 ft.   **(b)** 15 ft.   **(c)** 20 ft.   **(d)** 25 ft.   **(e)** 30 ft.

14. Emily is traveling in London. She can exchange $5 for 3 British pounds (£). How many pounds (£) will she receive for $65?
    (a) 29 £   **(b)** 39 £   **(c)** 49 £   **(d)** 19 £   **(e)** 37 £

15. On a tour guide map of Canada, 2 inches on the map represents 260 miles. How many miles does 3 inches represent?
    (a) 290 mi.   **(b)** 390 mi.   **(c)** 410 mi.   **(d)** 370 mi.   **(e)** 310 mi.

16. If Wendy pedals her bicycle at 84 revolutions per minute, she travels at 14 miles per hour. How fast does she go if she pedals at 96 revolutions per minute?
    (a) 26 mph   **(b)** 16 mph   **(c)** 24 mph   **(d)** 28 mph   **(e)** 22 mph

17. A bottle of spurge and oxalis killer for your lawn states that you need to use 2 tablespoons to treat 300 square feet of lawn. How many tablespoons will you need to use to treat 1500 square feet of lawn?
    (a) 8   **(b)** 10   **(c)** 12   **(d)** 14   **(e)** 16

18. Devon has a small cement patio 5 feet wide and 7 feet long in his yard. He wants to enlarge the patio, keeping the dimensions of the new patio proportional to those of the old patio. If he has room to increase the length to 21 feet, how wide should the patio be?
    (a) 18 ft.   **(b)** 15 ft.   **(c)** 20 ft.   **(d)** 25 ft.   **(e)** 24 ft.

Use the following information to answer questions 19 and 20. Two partners, John Ling and Kelvey Marks, each invest money in their business at a ratio of 6 to 7, with Kelvey investing the larger amount.

19. If John invested $2400, how much did Kelvey invest?
 (a) $2000　　**(b)** $2200　　**(c)** $2800　　**(d)** $2600　　**(e)** $3000

20. If the profits from the partnership are distributed to John and Kelvey based on the ratio of their investment, how much profit will John receive if Kelvey receives $798 for profits?
 (a) $784　　**(b)** $884　　**(c)** $684　　**(d)** $920　　**(e)** $690

21. Koursh conducted a science experiment and found that sound travels 34,720 feet in air and 31 seconds. How many feet would sound travel in 50 seconds?
 (a) 54,000 ft　**(b)** 26,000 ft　**(c)** 56,000 ft　**(d)** 26,500 ft　**(e)** 36,000 ft

22. In a small city located in the Midwest, 62 out of every 100 registered voters cast a vote in the last election. If there were 22,550 registered voters, how many people voted?
 (a) 45,100　　**(b)** 22,500　　**(c)** 10,981　　**(d)** 13,950　　**(e)** 13,981

23. Natalie increased her wealth by making very profitable investment choices. She told other investors that for every $500 she invested, she earned $800. How much did she invest to earn $1 million?
 (a) $500,000　**(b)** $450,000　**(c)** $560,000　**(d)** $625,000　**(e)** $734,000

24. If the property tax is $1600 on a home valued at $256,000, how much will the property tax be on a home valued at $192,000?
 (a) $1000　　**(b)** $1100　　**(c)** $1200　　**(d)** $1300　　**(e)** $1400

25. Jason jogged 8 miles yesterday and 2 miles today. The total distance took him 4 hours. If Barbara rides her bike at the same speed as Jason jogs, how long will it take her to travel 25 miles?
 (a) 5 hours　　**(b)** 1 hour　　**(c)** 10 hours　　**(d)** 2.5 hours　**(e)** 3 hours

Use the following information to answer questions 26 28. A motorcycle travels 50 miles on 1 gallon of gasoline at a speed of 40 mph.

26. What is the rate of gasoline consumption?
 (a) 40 mpg　　**(b)** 50 mpg　　**(c)** 30 mpg　　**(d)** 20 mpg　　**(e)** 60 mpg

27. How many miles are traveled if the motorcycle drives for 10 hours at 40 mph?
 (a) 500 mi.　**(b)** 400 mi.　**(c)** 300 mi.　**(d)** 200 mi.　**(e)** 150 mi.

28. How many gallons are needed to travel for 10 hours at 40 mph?
  (a) 6 gal   **(b)** 5 gal   **(c)** 8 gal   **(d)** 10 gal   **(e)** 50 gal

29. A 189 square foot rectangular floor is covered by square ceramic tiles. The width of the floor is spanned by 7 tiles, the length by 9 tiles. Given the same tile size, how large a floor in square feet, would be covered by 42 of these tiles?
  (a) 191 sq ft   **(b)** 181 sq ft   **(c)** 171 sq ft   **(d)** 126 sq ft   **(e)** 116 sq ft

30. John paid $37 for 5 CDs. At that rate what would 11 CDs cost?
  (a) $7.40   **(b)** $74.00   **(c)** $84.70   **(d)** $81.40   **(e)** $75.60

31. If $1\frac{1}{2}$ pounds of cherries cost $2.70, how much would $\frac{2}{3}$ pounds cost at the same rate?
  (a) $1.50   **(b)** $1.66   **(c)** $1.20   **(d)** $2.50   **(e)** $2.67

32. The dosage of a certain medicine is 3 ounces for every 60 pounds of body weight. How many ounces are needed for a person weighing 195 pounds?
  (a) 10.5 oz   **(b)** 9.5 oz   **(c)** 10.75 oz   **(d)** 9.75 oz   **(e)** 7.95 oz

33. A solution contains 40 gallons of fluids, which is a mixture of 3 parts chlorine to 2 parts water by volume. How many gallons of chlorine does the solution contain?
  (a) 20 gal   **(b)** 22 gal   **(c)** 24 gal   **(d)** 30 gal   **(e)** 32 gal

34. The cost of tiling a floor is proportional to the area of the floor. It costs $125 to tile a rectangular floor that is 10 feet long and 6 feet wide. What should the approximate cost be for the tiling of a rectangular floor that is 260 square feet in area?
  (a) $555   **(b)** $542   **(c)** $572   **(d)** $532   **(e)** $511

35. If 24 ounces of a certain liquid fills $\frac{1}{4}$ of a pail, how many ounces of the same liquid will fill $\frac{1}{3}$ of the pail?
  (a) $\frac{33}{3}$ **(b)** 18   **(c)** 30   **(d)** 32   **(e)** 36

36. Three-fifths of a class are girls. How many students are there in the class if there are 75 girls in the class?
  (a) 50   **(b)** 75   **(c)** 100   **(d)** 125   **(e)** 120

Chapter 6

Percentages

Percents are used in business (for various types of loan) in science (to measure the percent concentration of an acid in a solution) in economics (to express increases and decreases of the various economic indices) and in many other areas of our everyday life.

SECTION A. Understanding the Meaning of Percent

Percent means "parts of 100." As we have seen in previous chapters, decimals and fractions were used to represent parts of a whole. So too percent will be used to describe parts of a whole (the whole now being 100).

For example, 25 percent, written 25%, means 25 hundredths, i.e. 25 out of 100. Thus, in a class of 100 students, if 25% of the students obtained A's, we mean 25 students out of the 100 students obtained an A. In general, a given number of percent is equivalent to a fraction whose numerator is the given number and whose denominator is 100. Therefore, $25\% = \dfrac{25}{100} = 0.25$.

Notice that if 25% of the class got A's, then 75 out of 100 students did not, i.e., 75% of the student did not get A's. The percent that got A's plus the percent that did not get A's: $25\% + 75\% = 100\%$, which represents the entire class.

From the preceding discussion, percent has the following meaning:
  a. as hundredth,
  b. as a fraction where denominator is 100,
  c. as a 2 place decimal,
  d. as a number out of 100.

EXAMPLE 1: State using percents: 13 out of 100 radios are defective.

SOLUTION: $\dfrac{13}{100} = 13\%$ 13% of the radios are defective.

EXAMPLE 2: Last year's attendance at the school's winter formal was 100 students. This year the attendance was 121. Write this year's attendance as a percent of last year's.

SOLUTION:

$$\text{This year's attendance} \rightarrow \frac{121}{100} = 121\%$$
$$\text{Last year's attendance} \rightarrow$$

This year's attendance at the formal was 121% of last year's.

We note that percents can be larger than 100% (as in the above example) or less than 1% (as in the next example).

EXAMPLE 3: There are 100 milliliters (mL) of solution in a container. Sara takes 0.3 mL of the solution. What percentage of the fluid does Sara take?

SOLUTION: Sara takes 0.3 mL out of 100 mL, or $\frac{0.3}{100} = 0.3\%$ of the solution.

EXAMPLE 4: In a mixture of alcohol and water, what percent of the mixture is alcohol if the percents water is 90%?

SOLUTION: Since 100% represents the entire mixture, if water is 90% then the percent alcohol is $100\% - 90\% = 10\%$.

EXAMPLE 5: Express each of the following as a percent.
    a. 30.2 hundredths
    b. $13\frac{1}{8}$ out of a hundred
    c. $\frac{45}{100}$
    d. 0.62

SOLUTION:
    a. 30.2%
    b. $13\frac{1}{8}\%$
    c. 45%
    d. $0.62 = \frac{62}{100} = 62\%$

194

Exercises: Understanding the Meaning of Percent

For 1 – 13, state using percents.

1. 31 out of 100 students in the class voted.

2. 7 out of 100 phones are defective.

3. 63 out of 100 power boats had a radar navigation system.

4. Last year's attendance at medical school was 100 students. This year the attendance is 113. Write this year's attendance as a percent of last year's.

5. 16 out of 100 people use electric toothbrushes. What percentage of people use electric toothbrushes?

6. 71 out of 100 students in the class are women.

7. 11 out of 100 new computers are defective.

8. 93 out of 100 new cars have compact disc players.

9. Last year's attendance at the College Service Club was 100 students. This year the attendance is 160. Write this year's attendance as a percent of last year's.

10. 58 out of 100 people play organized sports. What percentage of people play organized sports?

11. 31 out of 100 students did not vote.

12. 7 out of 100 phones are not defective.

13. For a mixture of water and alcohol, what percent of the mixture is water if the percent alcohol is 42%?

For 14 – 18, express each as a percent.

14. 113 hundredths

15. $14\frac{1}{2}$ out of a hundred

16. $\frac{63}{100}$

17. 0.84

18. 45 hundredths

19. What percent of a class is present if the percent absent is 70%?
    (a) 30%　　　**(b)** 70%　　　**(c)** 10%　　　**(d)** 60%　　　**(e)** 60%

20. Which of the following is not equivalent to 17%?
    (a) 0.17　　　　　　　　**(b)** $\frac{17}{100}$　　　　　　　　**(c)** 17 hundredths
    (d) 83%　　　　　　　　**(e)** 17 out of one hundred

SECTION B. Converting Decimal to Percent and Percent to Decimal

In many applications involving percents, it is necessary to convert percents to an equivalent decimal and a decimal to an equivalent percent.

To convert a percent to an equivalent decimal
1. Remove the percent sign.
2. Multiply by 0.01, i.e. move the decimal point 2 places to the left.

$$14\% \ = \ 14 \times 0.01 \ = \ 0.14$$

| Remove % sign. |
| Move decimal point |
| 2 places to the left. |

EXAMPLE 1:   Write as a decimal.
    a. 3.82%   **b.** 280%   **c.** 0.6%

SOLUTION:
    a. $3.82\% = 3.82 \times 0.01 = 0.0382$
    b. $280\% = 280 \times 0.01 = 2.80$
    c. $0.6\% = 0.6 \times 0.01 = 0.006$

To convert a decimal to an equivalent percent
1. Multiply by 100, i.e. move the decimal point 2 places to the right.
2. Write the % sign at the end of the number.

$$0.16 : \quad 0.16 \times 100 \ = \ 16\%$$

| Move decimal point |
| 2 places to the right. |
| Add the % sign. |

EXAMPLE 2:   Write as a percent.
    a. 0.613   **b.** 0.08   **c.** 2.81   **d.** 0.9

SOLUTION:
    a. $0.613: \ 0.613 \times 100 = 61.3\%$
    b. $0.08: \ 0.08 \times 100 = 8\%$
    c. $2.81: \ 2.81 \times 100 = 281\%$
    d. $0.9: \ 0.9 \times 100 = 90\%$ We insert a zero after the 9.

Exercises: Converting Decimal to Percent and Percent to Decimal

1. Write 53.8% as a decimal.

2. Write 0.0024 as a percent.

3. Write 2.33% as a decimal.

4. In Alaska, 0.03413 of the state is covered by water. Write the part of the state that is covered by water as a percent.

5. Write 24.4% as a decimal.

6. Write 0.006 as a percent.

7. Write 5.2% as a decimal.

8. In Florida, 0.07689 of the state is covered by water. Write the part of the state that is covered by water as a percent.

For 9 – 10, write each decimal as a percent.
9. 0.476   10. 11.88

For 11 – 12, write each percent as a decimal.
11. 1312%   12. 0.6%

For 13 – 16, write each decimal as a percent.
13. 0.05   14. 0.9

15. 0.52   16. 0.92

For 17 – 20, write each percent as a decimal.
17. 5%   18. 86%

19. 105%   20. 30.6%

SECTION C. Converting a Fraction to a Decimal and a Decimal to a Fraction

It is often necessary to convert a decimal to an equivalent fraction and a fraction to an equivalent decimal.

To write a fraction as a decimal
1. Divide the numerator by the denominator.
2. The quotient can be rounded to the desired number of decimal places.

EXAMPLE 1:   Convert to a decimal (round to 3 decimal positions)

   a. $\dfrac{1}{2}$ **b.** $\dfrac{1}{4}$ **c.** $\dfrac{7}{11}$ **d.** $\dfrac{1}{7}$

SOLUTION:

a. $\dfrac{1}{2}$ $\xrightarrow{\text{divide}}$ $2\overline{)1.0}$ with quotient 0.5

$$\begin{array}{r} 0.5 \\ 2\overline{)1.0} \\ \underline{10} \\ 0 \end{array}$$ ← remainder zero
$\qquad$ Thus $\dfrac{1}{2} = 0.5$

b. $\dfrac{1}{4}$ $\xrightarrow{\text{divide}}$

$$\begin{array}{r} 0.25 \\ 4\overline{)1.00} \\ \underline{8} \\ 20 \\ \underline{20} \\ 0 \end{array}$$ ← remainder zero
$\qquad$ Thus $\dfrac{1}{4} = 0.25$

c. $\dfrac{7}{11}$ $\xrightarrow{\text{divide}}$

$$\begin{array}{r} 0.6363 \\ 11\overline{)7.0000} \\ \underline{66} \\ 40 \\ \underline{33} \\ 70 \\ \underline{66} \\ 40 \\ \underline{33} \end{array}$$

We note that this pattern repeats.

d. $\dfrac{1}{7}$ $\xrightarrow{\text{divide}}$

$$\begin{array}{r} 0.1428 \\ 7\overline{)1.0000} \\ \underline{7} \\ 30 \\ \underline{28} \\ 20 \\ \underline{14} \\ 60 \\ \underline{56} \\ 4 \end{array}$$
$\qquad$ Thus $\dfrac{1}{7} = 0.143$

Note: division takes place until
    a. the remainder is zero (as in example 1a)
    b. the reminder repeats itself (as in example 1c)
    c. the desired number of decimal places has been achieved (as in example 1d).

When a decimal has a digit or a group of digits that repeat (as in example 1c), we indicate the repeating digits or group of digits with a bar over the repeating digit or group of repeating digits. Thus $\frac{7}{11} = 0.\overline{63} = 0.636363\ldots$

EXAMPLE 2: Convert $\frac{4}{7}$ to a decimal.

SOLUTION: $\frac{4}{7} = 7 \overline{)\begin{array}{l} 0.57142 \\ 4.00000 \end{array}}$

$$\begin{array}{r} \underline{35} \\ 50 \\ \underline{49} \\ 10 \\ \underline{7} \\ 30 \\ \underline{28} \\ 20 \\ \underline{14} \\ 6 \end{array}$$

Thus $\frac{4}{7} = 0.6$ rounded to the nearest tenth

      $= 0.57$ rounded to the nearest hundredth

      $= 0.571$ rounded to the nearest thousandth

      $= 0.5714$ rounded to the nearest ten-thousandth

To convert a decimal to a fraction
1. Write the whole number (the digits to the left of the decimal point).
2. Write the decimal part (the digits to the right of the decimal point) over a denominator that has a 1 and the same number of zeros as the number of digits in the decimal part.

EXAMPLE 3: Convert to a fraction.
      a. 0.62   **b.** 4.63

SOLUTION:

    a. $0.62 \longrightarrow 0$ Step 1 write the whole number.

$$\frac{62}{100}$$ Step 2 decimal part 62 over denominator 100 (2 zeros

                           for 2 decimal positions.

    Thus $0.62 = \dfrac{62}{100}$

    b. 4.63   4   Step 1

                  $\dfrac{63}{100}$ Step 2

    Thus $4.63 = 4\dfrac{63}{100}$ 2 zeros correspond to two decimal places.

SECTION D. Changing between Fractions, Decimals and Percents

Now that you can change between decimals and percents, and between decimals and fractions, you are ready to change between fractions and decimals and percent.

$$\text{fraction} \underset{\rightarrow}{\overset{\leftarrow}{}} \text{decimal} \underset{\rightarrow}{\overset{\leftarrow}{}} \text{percent}$$

To change a fraction to a percent
1. Convert the fraction to an equivalent decimal.
2. Convert the decimal to a percent.

EXAMPLE 1: Convert to a percent.

    a. $\dfrac{211}{250}$ **b.** $\dfrac{7}{9}$ (round to the nearest hundredth)

SOLUTION:

    a. $\dfrac{211}{250} = 0.844$ Convert fraction to decimal by dividing

            $0.844 = 84.4\%$ Convert decimal to percent by moving decimal point 2 positions to the right

    b. $\dfrac{7}{9} = 0.\overline{7} = 0.77777\ldots$ Convert to decimal

        $0.77777 = 77.78\%$ Convert to percent

Note: if converting a fraction to a decimal results in a repeating decimal we usually carry out the division at least five places past the decimal point so that when we convert to % (moving the decimal places 2 places to the right) we will be left with 3 decimal positions to enable us to round to the nearest hundredth (as in example 1b).

You can convert a fraction to a percent by simply multiplying the fraction by 100 and dividing (and appending the % sign at the end).

EXAMPLE 2:　Convert to percent.

　　a. $\frac{211}{250}$　　　　　　　　**b.** $\frac{7}{9}$　(round to the nearest hundredth)

SOLUTION:

　　a.　$\frac{211}{250} \times 100 = \frac{21,100}{250} = \frac{2110}{25}$

$$25\overline{)2110.0}$$
$$\underline{200}$$
$$110$$
$$\underline{100}$$
$$100$$
$$\underline{100}$$
$$0$$

quotient 84.4 → 84.4%

　　b.　$\frac{7}{9} \times 100 = \frac{700}{9}$

$$9\overline{)700.000}$$
$$\underline{63}$$
$$70$$
$$\underline{63}$$
$$70$$
$$\underline{63}$$
$$70$$
$$\underline{63}$$
$$70$$
$$\underline{63}$$
$$7$$

quotient 77.777 → 77.78%

To write a percent as a fraction
1. Convert the percent to an equivalent decimal.
2. Convert the decimal to a fraction.

EXAMPLE 3:　Write as a fraction.

　　a. 42.1%　　　　　　　**b.** $\frac{1}{4}\%$

SOLUTION:

　　a.　$42.1\% = 0.421$　　Step 1 Convert to decimal. Move decimal point 2 places to the left.

　　　　$0.421 = \frac{421}{1000}$　　Step 2 Convert to fraction. Divide decimal part by 1000 (1 followed by 3 zeros since 3 is the number of decimal positions.)

　　b.　$\frac{1}{4}\% = 0.25\% = 0.0025$　　Convert $\frac{1}{4}$ to decimal and convert % to decimal

　　　　$0.0025 = \frac{25}{10,000} = \frac{1}{400}$　　Divide decimal part 0.0025 by 1 followed by 4 zeros

　　　　　　　　　　　(4 is the number of decimal positions) and we then simplify.

203

EXAMPLE 4:

 a. Write $\dfrac{35}{40}$ as a percent.

 b. Write $\dfrac{3}{4}\%$ as a fraction.

SOLUTION:

 a. $\dfrac{35}{40} = 0.875$ Divide 35 by 40 to get 0.875.

 $0.875 \rightarrow 87.5\%$ Convert to percent by shifting decimal point 2 places to the right.

 Thus $\dfrac{35}{40} = 87.5\%$

 b. $\dfrac{3}{4}\% = 0.75\%$ Convert fraction $\dfrac{3}{4}$ to decimal $\dfrac{3}{4} = 0.75$.

 $0.75\% \rightarrow 0.0075$ Convert percent to decimal by shifting decimal point 2 places to the left.

 $0.0075 = \dfrac{75}{10,000} = \dfrac{3}{400}$ Convert decimal to fractions.

Certain percents occur very often, especially in money matters. Here are some common equivalents that you may already know. If not, be sure to get to know them.

$$\dfrac{1}{4} = 0.25 = 25\% \qquad \dfrac{1}{3} = 0.3\overline{3} = 33\dfrac{1}{3}\% \qquad \dfrac{1}{10} = 0.10 = 10\%$$

$$\dfrac{1}{2} = 0.5 = 50\% \qquad \dfrac{2}{3} = 0.6\overline{6} = 66\dfrac{2}{3}\% \qquad \dfrac{3}{4} = 0.75 = 75\%$$

EXAMPLE 5: Convert $133.\overline{3}\%$ to a fraction.

SOLUTION: $133.\overline{3}\% = 1.33\overline{3} = 1\dfrac{1}{3}$ Since $0.3\overline{3} = 0.33\overline{3} = 0.\overline{3} = \dfrac{1}{3}$

$\phantom{133.\overline{3}\% } = \dfrac{4}{3}$ Convert to improper fraction (if needed)

Exercises: Changing between Fractions, Decimals and Percents

1. Write $\dfrac{129}{250}$ as a percent.

   (a) 129%   (b) 250%   (c) 48%   (d) 51.6%   (e) 61%

2. Write $\dfrac{1}{2}$% as a fraction.

   (a) 0.005   (b) $\dfrac{1}{200}$ (c) 0.5%   (d) 50%   (e) 25%

3. Write $\dfrac{14}{40}$ as a percent.

   (a) 14%   (b) 40%   (c) 35%   (d) 30%   (e) 28%

4. Write 22.3% as a fraction.

   (a) $\dfrac{223}{1000}$ (b) $\dfrac{1}{22.3}$ (c) $\dfrac{22.3}{1000}$ (d) 223%   (e) $\dfrac{223}{100}$

5. Write $\dfrac{32}{80}$ as a percent.

   (a) 0.4   (b) 4%   (c) 40%   (d) 32%   (e) 80%

6. Write 72.1% as a fraction.

   (a) $\dfrac{721}{100}$ (b) $\dfrac{721}{1000}$ (c) $\dfrac{721}{10,000}$ (d) $\dfrac{1}{74}$ (e) not given

7. Arran Copy Center wastes $2\dfrac{1}{2}$% of its paper supply due to poor quality of the photocopies produced. Write this percent as a fraction.

   (a) $\dfrac{1}{40}$ (b) $\dfrac{2.5}{1000}$ (c) $\dfrac{2.5}{40}$ (d) $\dfrac{5}{2}$ (e) not given

8. During waking hours, a person blinks $\dfrac{9}{2000}$ of the time. Express the fraction as a percent.

(a) 4.5%   (b) 0.45%   (c) 45%   (d) 209%   (e) 191%

9. A seamstress wastes $1\dfrac{1}{4}$% of the material used to make a dress. Write this percent as a fraction.

(a) $\dfrac{5}{4}$ (b) $\dfrac{1}{8}$ (c) $\dfrac{1}{80}$ (d) 1.25   (e) $\dfrac{5}{40}$

10. The brain represents $\dfrac{1}{40}$ of an average person's weight. Express this fraction as a percent.

(a) 2.5%   (b) 25%   (c) 0.25%   (d) 0.40%   (e) 4.0

For 11 – 13, write each percent as a fraction.

11. 80%

(a) $\dfrac{8}{100}$ (b) $\dfrac{4}{5}$ (c) $\dfrac{4}{100}$ (d) $\dfrac{80}{20}$ (e) not given

12. $433.\overline{3}$%

(a) $\dfrac{43.33}{100}$ (b) $\dfrac{4333}{10}$ (c) $\dfrac{4.33\overline{3}}{10}$ (d) $4.3\overline{3}$ (e) $4\dfrac{1}{3}$

13. $33\dfrac{1}{3}$%

(a) $\dfrac{1}{3}$ (b) $0.33\overline{3}$ (c) $\dfrac{1}{6}$ (d) $\dfrac{3}{100}$ (e) not given

206

For 14 – 20, write each fraction as a percent.

14. $\dfrac{20}{50}$

    (a) 20%   **(b)** 50%   **(c)** 30%   **(d)** 40%   **(e)** not given

15. $\dfrac{7}{1000}$

    (a) 0.7%   **(b)** 7%   **(c)** 70%   **(d)** 700%   **(e)** not given

16. $5\dfrac{1}{2}$

    (a) 50%   **(b)** 55%   **(c)** 5.5%   **(d)** 550%   **(e)** not given

17. $\dfrac{7}{250}$

    (a) 280%   **(b)** 28%   **(c)** 7%   **(d)** 2.8%   **(e)** not given

18. $\dfrac{1}{8}$

    (a) 8%   **(b)** 11%   **(c)** 12%   **(d)** 2.4%   **(e)** not given

19. $\dfrac{1}{3}$

    (a) $33\dfrac{1}{3}\%$ **(b)** $66\dfrac{2}{3}\%$ **(c)** 3%   **(d)** 30%   **(e)** not given

20. $\dfrac{2}{3}$

    (a) $33\dfrac{1}{3}\%$ **(b)** $66\dfrac{2}{3}\%$ **(c)** 3%   **(d)** 30%   **(e)** not given

SECTION E. Setting Up and Solving Percent Problems

Percents are used to describe parts of a whole base amount. For example, 80% describes the amount of *4 parts out of 5.*

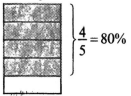

$$\frac{4}{5} = 80\%$$

We see that 4 parts of the whole (5 parts) is 80% of the whole (also called base), i.e. 80% of 5 is 4. In general, we have the *basic percent relationship* or *basic percent equation*

Percent × base = amount (this is the part being compared to the base).

    80% × 5 = 4

When two of these 3 quantities are known we can use this basic relationship to find the third quantity. In problems when we perform computation with percents we must first convert the percent into the equivalent decimal or fraction before the basic relationship is solved for the unknown quantity.

When we set up the problem, we are translating the given problem or sentence into mathematical symbols. When this translating takes place, the word "of" is written as × (times), "is" is written as = (equals), and "what" as well as the word "find" is written as x or n (the unknown number).

The basic percent relationship can be formulated into three basic ways – these are called *basic percent statements.*

1. What is (or find) $a\%$ of b? Here we are given a and b (the base). We want to find the amount.

    amount = a (in decimal) × b

2. What percent of a is b (or equivalently b is what percent of a?) Here we are given the amount and base, we want to find the %.

    $\% = \dfrac{\text{base}}{\text{amount}}$ (after division we convert quotient to %).

3. a is b of what number. Here we know the amount and % we want to find the base.

    $\text{base} = \dfrac{\text{amount}}{\% \text{ (in decimal)}}.$

EXAMPLE 1:   Set up the basic percent relationship and solve.

    (a) What is 25% of 40?

    (b) 10 is 25% of what number?

    (c) 10 is what percent of 40?

SOLUTION: We translate each into the form $\text{amount} = \text{percent} \times \text{base}$.

    (a) What is 25% of 40?

$$x \quad = \quad 25\% \times \quad 40 \qquad \text{Write in symbols.}$$

    We want to find the *amount*, that is, the *part* of (25% of) the *base* of 40.

$$x = 25\% \text{ of } 40$$

$$x = 0.25 \times 40 \qquad \text{Change 25\% to decimal form.}$$

$$x = 10 \qquad\qquad\quad \text{Multiply.}$$

  10 is 25% of 40.

    (b) 10 is 25% of what number?

$$10 = 25\% \times \quad x$$

    We want to find the *base* (the entire quantity).

$$10 = 0.25 \times x \qquad \text{Change 25\% to a decimal.}$$

$$\frac{10}{0.25} = x \qquad\qquad \text{Solve the equation for } x.$$

$$40 = x \qquad\qquad\quad \text{Divide.}$$

  10 is 25% of **40**.

    (c) 10 is what percent of 40?

$$10 = \quad x \qquad \% \qquad \times \quad 40$$

    We want to find the *percent*.

$$10 = x\% \times 40$$

$$\frac{10}{40} = x\% \qquad\qquad \text{Solve the equation for } x\%.$$

$$0.25 = x\% \qquad\qquad \text{The \% symbol reminds us to write 0.25 as a percent.}$$

$$25\% = x \qquad\qquad \text{Change 0.25 to a percent.}$$

  10 is **25%** of 40.

    Note that the answer must be in percent form because the question asks <u>what percent</u>.

Sometimes we have more than 100% - this means that the *amount* is more than the *base* amount.

EXAMPLE 2:   Set up the basic percent relationship and solve. 50 is what percent of 40?

SOLUTION: We should expect to get more than 100% since 50 is more than the base 40.

50   is what percent of 40?

↓ ↓ ↓   ↓   ↓   ↓

50 $=$ x % $\times$ 40

50 $= x\% \times 40$

$\dfrac{50}{40} = x\%$ Solve for $x\%$.

$1.25 = x\%$ Divide.

$125\% = x$ Write 1.25 as a percent.

50 is 125% of 40.

Exercises: Setting Up and Solving Percent Problems

1. What is 32% of 84?
    (a) 26.88   (b) 42.86   (c) 22.88   (d) 25.26   (e) not given

2. Find 53% of 210.
    (a) 50.30   (b) 111.3   (c) 70.9   (d) 80.6   (e) not given

3. What is 26% of 72?
    (a) 18.50   (b) 18.72   (c) 20   (d) 70   (e) not given

4. What is 46% of 60?
    (a) 25.8   (b) 28.2   (c) 27.6   (d) 32   (e) not given

5. Find 24% of 145.
    (a) 20.6   (b) 34.8   (c) 36.2   (d) 36.9   (e) not given

6. What is 18% of 66?
    (a) 11.88   (b) 12.80   (c) 13.72   (d) 12.52   (e) not given

7. What percent of 650 is 70?
    (a) 9.58%   (b) 15.34%   (c) 10.77%   (d) 8.16%   (e) 23.53%

8. 60 is what percent of 30?
    (a) 200%   (b) 2%   (c) 20%   (d) 100%   (e) 60%

9. What percent of 350 is 20?
    (a) 18.6%   (b) 5.7%   (c) 23.6%   (d) 20%   (e) 350%

10. 400 is what percent of 80?
    (a) 5%   (b) 50%   (c) 500%   (d) 0.5%   (e) 0.05%

11. 56 is 70% of what number?
    (a) 80   (b) 85   (c) 90   (d) 95   (e) 100

12. 24 is 40% of what number?
(a) 50   (b) 60   (c) 70   (d) 80   (e) 90

13. 72 is 25% of what number?
(a) 280   (b) 288   (c) 290   (d) 270   (e) 275

14. 32 is 64% of what number?
(a) 40   (b) 50   (c) 60   (d) 70   (e) 80

15. 2.8 is what percent of 3.5?
(a) 125%   (b) 90%   (c) 80%   (d) 20%   (e) 60%

16. What is 250% of 300?
(a) 750   (b) 550   (c) 7500   (d) 75,000   (e) 75

17. Which of the following is closest to 28% of 655?
(a) 183   (b) 193   (c) 195   (d) 200   (e) 180

18. To find 36% of a number you would
(a) multiply the number by 36
(b) multiply the number by 0.36
(c) divide the number by 36
(d) divide the number by 0.36
(e) divide the number by 3.6

19. What percent of 40 is 25?
(a) 62.5%   (b) 160%   (c) 85%   (d) 6.25%   (e) 55%

20. 430 is 40% of what number?
(a) 172   (b) 1075   (c) 1720   (d) 17,200   (e) 107.5

21. If 3% of a number is 9, than that number is
(a) 30   (b) 300   (c) 3   (d) 3000   (e) 0.3

22. To find 300% of a number, we would multiply that number by
(a) 30   (b) 300   (c) 3   (d) 0.3   (e) 0.03

SECTION F. Solving Applied Problems Involving Percent

We can solve word problems that involve percents by using the following 3-step process.

1. Writing a percent statement to represent the situation.
2. Translating the statement to a basic percent equation.
3. Solving the equation.

EXAMPLE 1: Marilyn has 850 out of 1000 points possible in her English class. What percent of the total points does Marilyn have?

SOLUTION: We must find the *percent*, so we write the percent statement that represents the percent situation.

  850 is what percent of 1000?   Write the statement that represents

  ↓ ↓   ↓   ↓   ↓   ↓   this situation.

  850 = x %   × 1000? Form a basic percent equation.

  $$\frac{850}{1000} = x\% \times \frac{1000}{1000}$$ Solve the equation.

  $0.85 = x\%$

  $85\% = x$

Marilyn has 85% of the total points.

We note the above basic percent statement is equivalent to "what percent of 1000 is 850?"

EXAMPLE 2: Sean's bill for his dinner at the Spaghetti House was $19.75. How much should he leave for a 15% tip? Round this amount to the nearest cent.

SOLUTION: We must find the *amount*, that is the *part* of (15% of) the *base* of $19.75.

  What is 15% of $19.75? Write the statement for this situation.

  ↓   ↓   ↓   ↓   ↓

  x = 15% × $19.75? Translate the statement into a basic percent equation.

  = 0.15   × $19.75

  = 2.9625   Solve the equation.

  = $2.96   Round.

The tip is $2.96.

213

EXAMPLE 3: Alex is a car salesman and earns a commission rate of 9% of the price of each car he sells. If he earned $3150 commission this month, what were his total sales for the month?

SOLUTION: amount = percent $\times$ base

Commission = commission rate $\times$ total sales

$$\$3150 = \quad 9\% \quad \times \quad x$$
$$\$3150 = 0.09 \times x$$
$$\frac{\$3150}{0.09} = \frac{0.09x}{0.09}$$
$$\$35,000 = x$$

Alex's total sales were $35,000.

Percent increase is used to indicate how much a quantity has increased over its original value. Similarly, percent decrease is used to show how much a quantity has decreased from its original value.

To find percent increase:
1. New value – original value = amount of increase
2. Percent increase $\times$ original value = amount of increase.

To find percent decrease:
1. Original value – new value = amount of decrease.
2. Percent decrease $\times$ original value = amount of decrease.

Note in either case we have our basic percent relationship:

Increase or decrease = percent increase or decrease $\times$ original values

AMOUNT   =   PERCENT $\times$ BASE

Once we find the amount of increase or decrease, we can find the new value by adding to or subtracting from the original value. Thus:

New value = original value + increase

New value = original value – decrease

214

EXAMPLE 4: The enrollment at Laird Elementary School was 450 students in 2000. In 2001, the enrollment decreased by 36 students. What was the percent decrease?

SOLUTION:

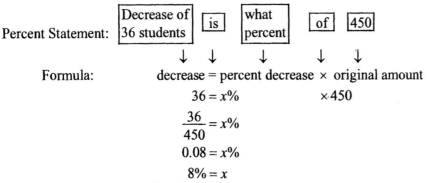

Percent Statement:

The enrollment decreased by 8%.

Note that we must change 0.08 to a percent since we are finding a percent.

EXAMPLE 5: Arnold earned $26,000 a year and received a 6% raise. How much is his new yearly salary?

SOLUTION: First, we find the amount of the raise.

     percent increase $\times$ original amount = increase (raise)

         6% $\times$ 26,000 $= 0.06 \times 26,000 = \$1560$ raise

Now we find his new yearly salary.

     original salary + raise = new salary

      $26,000   + \$1560 = \$27,560$

Arnold's new salary is $27,560.

215

EXAMPLE 6: An advertisement states that all items in a department store are reduced 30% off the original list price. What is the sale price of a big-screen television set with a list price of $2700?

SOLUTION: First, we find the amount of the discount.

percent decrease × amount – decrease (discount)

30% × 2700 $= 0.30 \times 2700 = \$810$ discount

Next we find the sale price.

original price – discount = sale

$2700 – $810 = $1890

The sale price is $1890.

EXAMPLE 7: At a sale in a clothing store, the sales price of a suit was $150.00. This price represented a 40% discount. What was the original price of the suit?

SOLUTION: The sales price was 60% of the original price, since there was a 40% discount. Therefore,

amount = percent × base (original amount)

150 = 0.60 × x

$\dfrac{150}{0.60} = \dfrac{0.60x}{0.60}$ Solve the equation.

$250 = x$

The original price of the suit was $250.

When you deposit money in a bank, the bank pays you rent for the privilege of using that money. The amount of rent to you is called **interest**. Similarly, when you borrow money from a bank, you pay for the privilege of using that money. The amount paid to the bank is also called **interest**. The original amount borrowed or deposited is called the **principal**. The amount of interest is a percent (called the **interest rate**) of the principal. Interest rates are given for specific periods of time, usually months or years. Interest computed on the original principal is called simple interest, which can be calculated using the following formula:

Simple interest = principal × rate × time

I = P × R × T

R is the rate per period of time, T is the number of time periods. The time period for R and T must be the same.

EXAMPLE 8: Larsen borrowed $9400 from the bank at a simple interest rate of 13% per year.
    (a) Find the interest on the loan for 1 year.
    (b) How much does Larsen pay back to the bank at the end of the year when he pays off the loan?

SOLUTION:
    (a) P = principal = $9400, R = rate = 13%, T = time period = 1 year.
$$I = P \times R \times T$$
$$I = \$9400 \times 0.13 \times 1 = \$1222 \quad \text{We change 13\% to 0.13, then multiply.}$$
    The interest for 1 year is $1222.
    (b) At the end of the year, he must pay back the original amount he borrowed plus the interest.
$$\text{original loan} + \text{interest} = \text{payoff amount}$$
$$\$9400 \quad + \quad \$1222 \ = \$10,\!622$$
    The total amount that Larsen must pay back is $10,622.

EXAMPLE 9: Find the interest on a loan of $2500 that is borrowed at a simple interest rate of 9% per year for 3 months.

SOLUTION: We must change 3 months to years since the formula requires that the time period for R and T to be the same: $T = 3$ months $= \dfrac{3}{12} = \dfrac{1}{4}$ year.
$$I = P \times R \times T$$
$$I = 2500 \times 0.09 \times \frac{1}{4} = 225 \times \frac{1}{4} = 56.25 \quad \text{We divide 225 by 4.}$$
The interest for 3 months is $56.25.

Sometimes we must solve a percent problem that depends on information obtained by solving another percent problem.

EXAMPLE 10: If 3% of a number is 9, what is 8% of that number?

SOLUTION:
$$\text{amount} = \text{percent} \times \text{base}$$
$$9 \ = \ 0.03 \ \times \ x$$
$$\frac{9}{0.03} = \frac{0.03x}{0.03} \qquad \text{Solve for } x.$$
$$300 = x \qquad \text{The number is 300.}$$
Now we find 8% of 300.
$$\text{amount} = 0.08 \times 300 = 24$$
Our answer is 24.

EXAMPLE 11: 60% of the graduating class are girls. Of the remaining graduates, 40% have blue eyes. What percentage of the graduating class are not girls and have blue eyes?

SOLUTION: 60% of the graduating class are girls. 40% are not. If we denote the number of students in the graduating class by x, then $0.40x$ is the number of students in the graduating class that are not girls. Therefore 40% of $0.40x$ represents the number of student graduates that are not girls but have blue eyes. Therefore:

$$(0.40)(0.40x) = 0.16x \text{ or } 16\% \text{ of } x$$

i.e., 16% of the entire graduating class are not girls and have blue eyes.

EXAMPLE 12: John, a student, takes a 3 hour exam. After an hour, John has answered 30% of the questions and has 10 correct and 5 incorrect answers. In order to answer 90% of all questions correctly, how many of the remaining questions must he answer correctly?

SOLUTION: He answered 15 questions and this represents 30% of the question. Thus, $15 = 0.30x$, where x is the total number of questions on the exam. Solving, we find that $x = 50$, i.e. there are 50 questions on the exam. To obtain 90% correct, he must answer $0.90(50) = 45$ questions correctly. He has so far answered 10 correctly. He must, therefore, answer correctly 35 of the remaining $50 - 15 = 35$ questions.

Sometimes we are asked to combine percentages of two groups.

EXAMPLE 13: In a large class of 200 students, 120 are girls, and 80 are boys. 90% of the girls passed the calculus exam, while 80% of the boys passed the calculus exam. What percent of the total number of students in the class passed the calculus exam?

SOLUTION: girls: $0.90 \times 120 = 108$ 108 girls passed the exam

boys: $0.80 \times 80 = 64$ 64 boys passed the exam

Therefore, $108 + 64 = 172$ students out of 200 passed the exam.

$\dfrac{172}{200} = 86\%$ 86% of the class passed the exam.

There are problems which require us to modify solutions (mixtures of two or more elements) to create specified concentrations of a particular element within the solution.

EXAMPLE 14: We want to create a mixture of gasoline and alcohol, which is to have 80% gasoline. How many gallons of alcohol must be added to 200 gallons of gasoline to make this mixture?

SOLUTION: After we add x gallons of alcohol, this entire mixture will contain 200 gallons of gasoline, and this represents 80% of this new mixture i.e., $(0.80) \times$ entire mixture $= 200$. So then the entire mixture contains $= \dfrac{200}{0.80} = 250$ gallons. Thus, we must add $250 - 200 = 50$ gallons of alcohol.

Exercises: Solving Applied Problems Using Percent

1. The bill for Marie's dinner was $12.95. How much should she leave for a 15% tip? Round your answer to the nearest cent.
    (a) $2.15   (b) $1.80   (c) $1.94   (d) $1.75   (e) $2.05

2. 60% of the graduates of Trinity, a two-year college, transfer to a four-year college. If the graduating class at Trinity has 650 students, how many are transferring to a four-year college?
    (a) 370   (b) 380   (c) 390   (d) 400   (e) 410

3. The Boyd farm is 250 acres. 70% of it is suitable land for farming. How many acres can be used to farm?
    (a) 160   (b) 165   (c) 170   (d) 175   (e) 180

4. H&B Manufacturing claims that no more than 0.5% of its parts are defective. If a client orders 8600 parts from H&B Manufacturing, what is the largest number of parts that could be defective?
    (a) 35   (b) 37   (c) 39   (d) 41   (e) 43

5. A snack bar has 80 calories. If 15 of those calories are from fat, what percent of the calories are from fat?
    (a) 17.92%   (b) 18.25%   (c) 18.50%   (d) 18.75%   (e) 19.15%

6. In a soccer game, Amy made 2 out of 5 shots on goal. What percent of shots did she make?
    (a) 30%   (b) 35%   (c) 40%   (d) 45%   (e) 50%

7. Wesley paid $21 tax when he bought a mountain bike for $300. What percent tax did he pay?
    (a) 4%   (b) 14%   (c) 7%   (d) 17%   (e) 21%

8. On the Almen High School basketball team, 9 out of the 22 players are over 6 feet tall. ·
    What percent of the players are over 6 feet tall?
    (a) 30%   (b) 60%   (c) 65%   (d) 32%   (e) not given

9. 5% of the employees at the Lido Insurance Company called in sick with the flu. If 10 employees called in sick, how many employees are there at the company?
    (a) 100   (b) 150   (c) 200   (d) 250   (e) 300

10. The new 8-mile nature trail is 125% of the length of the original trail. How long was the original trail?
(a) 6.0 mi.   (b) 6.2 mi.   (c) 6.4 mi.   (d) 6.6 mi.   (e) 6.8 mi.

11. 25% of the graduating class at Springdale Community College received a scholarship. If 1800 students received a scholarship, how many students are at the college?
(a) 7200   (b) 7225   (c) 7250   (d) 2750   (e) 3750

12. The new 350-seat auditorium contains 140% of the number of seats in the original auditorium. How many seats were in the original auditorium?
(a) 200   (b) 225   (c) 250   (d) 275   (e) 375

13. In a certain college, 180 students graduated. This number is exactly 25% of the total number of the students in the college. How many students are in the college?
(a) 450   (b) 720   (c) 4500   (d) 7200   (e) 3600

14. A large piece of land has an area of 16 square miles. On this land there is a lake that is one square mile in area. What percent of the piece of land is the lake?
(a) 62.5%   (b) 6.25%   (c) 12.5%   (d) 75%   (e) 60%

15. Mike received a 8% increase in his weekly salary. If his original salary was $350.00 per week, what was his new salary?
(a) $28   (b) $378   (c) $428   (d) $354.40   (e) $322

16. John makes $410 per week. How much would he make per week if he received a 9% raise?
(a) $506.40   (b) $450.75   (c) $478.65   (d) $446.90   (e) $419.00

17. A store buys 100 televisions for a total of $40,000. To determine the selling price it was decided to add a markup of 30% of the cost. What is the selling price of each television set?
(a) $400   (b) $520   (c) $650   (d) $600   (e) $430

18. Bob, a real estate agent, makes a commission of 6% of the selling price of every house he sells. How much commission does he make on a house that he sells for $350,000?
(a) $1800   (b) $21,000   (c) $6000   (d) $23,000   (e) $2100

19. A portion of pasta contains 200 calories, 10 of which are from fat. What percent of the calories in a portion of pasta are from fat?
(a) 5.25%   (b) 5%   (c) 20%   (d) 10%   (e) 6%

20. Daniel, a well known estate lawyer, had \$100,000 in a savings account that pays $4\frac{1}{2}\%$ simple interest. The money had been in the account for only six months when Daniel withdrew the money. He received $\frac{1}{2}$ of a year $4\frac{1}{2}\%$ simple interest. What was the total, savings plus interest, he had when he withdrew his money?
(a) \$104,500   (b) \$52,250   (c) \$102,250   (d) \$51,125   (e) not given

21. An appliance store is currently selling a certain TV set for \$550. Next month the price is expected to increase by 9%. Which expression can be used to calculate the new selling price?
(a) $(550)(0.09)$ (b) $550(0.09)+9$ (c) $550+550(0.09)$
(d) $550+550(9)$ (e) not given

22. The daily value for vitamin B-6 is 2 mg (milligrams). A capsule containing 25 mg of vitamin B-6 provides what percent of the daily value?
(a) 125%   (b) 12.5%   (c) 1250%   (d) 50%   (e) 8%

23. At a storewide sale, a store offers the following pricing policy. Each item will sell for 40% off the original price or for the reduced price already shown on the price tag, whichever is less. What will be the sales price of a dress that has already been reduced from the original price \$150.00 to the current price tag, \$100.00?
(a) \$90.00   (b) \$100.00   (c) \$40.00   (d) \$50.00   (e) \$95.00

24. Frank allocated \$300 of his weekly pay check for rent and food. He spends 20% of that \$300 per week for rent. Which of the following represents the amount of the \$300 that he has to spend on food in a 4-week period?
(a) $300(0.80)$ (b) $1200(0.80)$ (c) $1200 \div 0.20$
(d) $300 \div 0.20$ (e) $1200 \div 0.80$

25. After a 45% discount is made, the price of a dress is \$52.25. What is the original price of the dress?
(a) \$97.25   (b) \$100.00   (c) \$95.00   (d) \$48.75   (e) \$105.25

26. If 20% of a number is 14, what is 80% of that number?
(a) 56   (b) 70   (c) 11.2   (d) 2.8   (e) 25

27. A school has 2000 students. 35% are girls of which 20% have brown eyes. How many girls have brown eyes?
(a) 140   (b) 1400   (c) 700   (d) 400   (e) 1100

28. An appliance store, originally priced a freezer at $450 increases its prices by 2% on February 2, 2004. On August 15, 2004, the store again increased the current price by 5% of the February price. How many dollars is the current price above the original price of $450.

(a) $481.95   (b) $31.95   (c) $31.50   (d) $481.50   (e) $457.00

29. 70% of a class scored at least 70 on a math test. Of the remaining class, 60% scored 60 to 69. What percent of the entire class score less than 60%.

(a) 6%   (b) 18%   (c) 12%   (d) 40%   (e) 30%

30. Pedro received his monthly paycheck of $2000. He paid his rent of $1000 and spent $400 on food. He put the remaining money in his savings account. What percentage of his paycheck did he put in his savings account?

(a) 30%   (b) 3%   (c) 50%   (d) 40%   (e) 20%

31. The price of a suit increased to $350. This represents 140% of the original price. Which of the following represents the original price?

(a) $\dfrac{350}{0.40}$ (b) 350×1.40 (c) $\dfrac{350}{1.40}$

(d) $350 + 0.40(350)$ (e) $350 + (1.40)(350)$

32. A dress was bought by a store for $20.00. The store increased their cost by 30% to arrive at the selling price. The store then had a clearance sale and reduced the price of the dress by 30%. What was the sale prices of the dress?

(a) $18.20   (b) $20.00   (c) $12.80   (d) $12.00   (e) $26

33. 40% of 30 equal 20% of what number?

(a) 12   (b) 30   (c) 18   (d) 20   (e) 60

34. Joe, a real estate agent, gets 10% commission. He agrees to give 30% of his commission to Jane if she recommends a buyer to him. She does recommend a buyer of a $20,000 piece of property. How much does Jane get?

(a) $1000   (b) $2000   (c) $600   (d) $300   (e) $500

35. A piece of fabric $2\frac{1}{2}$ feet long, shrinks to $2\frac{1}{4}$ feet in a wash. What percent of the length of the original piece is lost when the fabric is washed?

(a) 10%   (b) 90%   (c) 20%   (d) 25%   (e) 22%

36. Mrs. Smith buys material and makes table cloths. Each table cloth will be $8\frac{1}{2}$ ft long.

She needs enough material for 8 such table cloths. She will add an additional 10% of the total for possible waste. How many feet of material, rounded to the nearest foot will she order?

    (a) 70   **(b)** 68   **(c)** 74   **(d)** 75   **(e)** 82

37. 70% of 180 boys have blue eyes and 55% of 160 girls have blue eyes. About what percent of the combined groups of 340 boys and girls have blue eyes?

    (a) 65%   **(b)** 86%   **(c)** 63%   **(d)** 56%   **(e)** 42%

38. At a recent party, 140 of the 270 invited guests come to the party before 9 P.M. By the time the party ended 90% of the invited guests showed up. How many guests came after 9 P.M.?

    (a) 100   **(b)** 243   **(c)** 103   **(d)** 140   **(e)** 29

39. Cindy bought a piece of land for $4000. The following year she sold it. Her profit was $300 plus 12% of her original cost. What percent of the original cost was her total profit?

    (a) 12%   **(b)** 13%   **(c)** 18.5%   **(d)** 19.5%   **(e)** 78%

40. A dress shop bought a dress for $150 and added 40% to its cost to arrive at its retail selling price. Two weeks later at a sale, the shop discounted the retail price by 20%. This new price was an increase of what percent of its original cost?

    (a) 12%   **(b)** 18%   **(c)** 40%   **(d)** 20%   **(e)** 10%

41. In 2000, the Home Sweet Home construction company built 1000 homes. In 2001, the number of homes built was 12% more than year before. The number of home built in 2002 was 5% less than that in 2001. How many homes did the company build in 2002?

    (a) 1120   **(b)** 56   **(c)** 1064   **(d)** 1176   **(e)** not given

42. Rent is 50% of Mrs. Cody's budget. If rent is the only item in the budget that went up, by what percentage will her total budget increase if her rent is increased by 30%?

    (a) 50%   **(b)** 40%   **(c)** 30%   **(d)** 20%   **(e)** 15%

43. Approximately how many gallons of alcohol must be added to 20 gallons of gasoline to make a gasoline-alcohol mixture that is 70% gasoline?

    (a) 8.6   **(b)** 14   **(c)** 6   **(d)** 9   **(e)** 8

44. There are 4000 students at a Midwestern college: 1500 freshmen, 1000 sophomores, 800 juniors, and 700 seniors. Linda is running for student council president. It is known that 60% of all students voted for Linda. If 70% of the freshmen, 50% of the sophomores, and 70% of the juniors votes, how many seniors voted for Linda?

(a) 390   (b) 400   (c) 410   (d) 290   (e) 300

45. The price of a shirt increased from $20 to $25. The increase was what percent of the original price?

(a) 5%   (b) 25%   (c) 20%   (d) $\frac{1}{4}$%   (e) 125%

46. The price of a $200 suit was reduced by 35%. A week later, the reduced price of this suit was increased by 35%. What is the price of the suit now?

(a) $200.00   (b) $170.00   (c) $175.50   (d) $130.00   (e) $210.00

47. The sales tax in a certain city increased from 4% to 5%. The increase is what percent of the original 4% sales tax?

(a) 20%   (b) 25%   (c) 1%   (d) 125%   (e) 15%

48. The price of a suit was $200 on January 15, 2001. The price increased by 5% on February 15, 2001. On June 1, 2001, the price again was increased by 5% of the February 15, 2001 price. By how much did the price increase on June 1, 2001?

(a) $10   (b) $20   (c) $10.50   (d) $20.50   (e) $220.50

49. George opened a savings account that pays $5\frac{1}{4}$% interest per year. George initially deposited $2000 into this account. After 6 months, George needed the money, so he withdrew his money. The bank would give him only $\frac{1}{2}$ of the year's $5\frac{1}{4}$% simple interest. What was the total in the account when George withdrew the funds?

(a) $2100   (b) $2150   (c) $1075   (d) $2105.50   (e) $2052.50

50. Mike owed $600 to a store which didn't change interest for the first month. The next month he made a payment of $100, but bought another $50 worth of merchandise on credit. At the end of the month he gets a bill from the store charging 3% interest on his entire unpaid balance. How much interest was Mike charged?

(a) $15.00   (b) $15.50   (c) $16.00   (d) $16.50   (e) not given

Chapter 7

Averages

SECTION A. Mean, Median, Mode

There are several types of averages. We will discuss three of the most common types.

1. The **mean** of a set of values is the sum of the values divided by the number of values. The mean is often called the **average**.

    EXAMPLE 1: Find the average or mean test score of a student who has test scores of 71, 83, 87 99, 80, and 90.

    SOLUTION: We take the sum of the six tests and divide the sum by 6.

    $$\text{Sum of test scores} \rightarrow \frac{71-83-87-99-80-90}{6} = \frac{510}{6} = 85$$
    $$\text{Number of tests} \rightarrow$$

    The mean is 85.

2. If a set of numbers is arranged in order from smallest to largest, the median is that value that has the same number of values above it as below it, i.e., the median is the middle value when the values are arranged in increasing order.

    EXAMPLE 2: Find the median of the following numbers: 14, 19, 1, 2, 4, 8, 7, 8, 23.

    SOLUTION: Arrange the numbers in increasing order: 1, 2, 4, 7, 8, 8, 14, 19, 23.

    The median is 8.
    $$\uparrow$$
    middle
    value

    If a list of number contains an even number of different items, then of course there is no one middle number. In this situation we obtain the median by taking the average of two middle numbers.

    EXAMPLE 3: Find the median of the following numbers: 13, 16, 18, 26, 31, 33, 38, and 39.

    SOLUTION: $\underbrace{13, 16, 18}_{\text{three numbers}}$ $\underbrace{26, 31}_{\substack{\text{two middle} \\ \text{numbers}}}$ $\underbrace{33, 38, 39}_{\text{three numbers}}$

    The average (mean) of 26 and 31 is $\dfrac{26+31}{2} = \dfrac{57}{2} = 28.5$

    Thus, the median value is 28.5.

Which one of the mean, median or mode should someone use to best indicate the center or average of a set of data? For example, suppose the salaries in the department are $10,000, $11,000, $14,000, $15,000, $19,000, and $21,000. All salaries are distributed fairly evenly, that is, there are no one value that is extreme (must larger or much smaller than the rest). The mean salary

$$\frac{10,000 + 11,000 + 14,000 + 15,000 + 19,000 + 210,000}{6} = 15,000$$

yields a fairly good idea of the "average" salary. The median is $14,500 with three salaries higher and three salaries lower. There is no mode. However, suppose the six employees in the department had salaries of $10,000, $11,000, $14,000, $15,000, $19,000 and $\underline{\$81,000}$. The mean is $25,000, the median is $14,500, and again there is no mode. The mean of $25,000 is high and not very revealing because the $81,000 is a relatively unusually high salary. The median of $14,500 with the three numbers above and the three numbers below gives the best indication of the center or average of the salaries in this set of data.

3. Another value that is sometimes used to describe a set of data is the **mode**. The *mode* of a set of data is the number or *numbers that occur most often*.
   If two values occur most often, we say that the data have two modes (*or* are **bimodal**).

EXAMPLE 4: Find the mode of each of the following.
a. A student's test scores of 89, 94, 96, 89, and 90.
b. The ages of students in a calculus class: 33, 27, 28, 28, 21, 19, 18, 25, 26, and 33.

SOLUTION:
a. The mode of 89, 94, 96, 89, and 90 is 89 since it occurs twice in the set of data.
b. The data 33, 27, 28, 28, 21, 19, 18, 25, 26, and 33 are bimodal since both 28 and 33 occur twice.

If there is no number that occurs more often than any other, we say that the set of numbers has no mode. For example, the set 1, 5, 7, 9, 11, 13 has no mode.

Exercises: Mean, Median Mode

For 1 – 6, find the mean:

1. A student received grades of 89, 92, 83, 96, and 99 on math quizzes.
    (a) 90   (b) 89   (c) 91.8   (d) 95   (e) 91.4

2. The Windy City Passport Photo Center received the following number of telephone calls over the last 6 days: 23, 45, 63, 34, 21, and 42.
    (a) 45   (b) 65   (c) 21   (d) 42   (e) not given

3. The last 5 houses built in town sold for the following prices: $189,000, $193,000, $162,000, $102, 000, and $189,000.
    (a) $165,000 (b) $167,000 (c) $170,000 (d) $180,000 (e) $169,000

4. A student received grades of 77, 84, 90, 92, 83, and 84 on history quizzes.
    (a) 84   (b) 85   (c) 86   (d) 85.2   (e) 85.4

5. The local Hertz rental car office received the following number of inquiries over the last 7 days: 34, 57, 61, 22, 43, 80, and 39.
    (a) 45   (b) 45.5   (c) 46   (d) 47   (e) not given

6. Luis priced a sofa at 6 local stores. The prices were $499, $359, $600, $450, $529, and $629.
    (a) 509   (b) 510   (c) 511   (d) 512   (e) not given

For 7 – 13, find the median:

7. The annual salaries of the employees of a local cable television office are $17,000, $11,600, $23,500, $15,700, $26,700, and $31,500.
    (a) $15,700 (b) $17,000 (c) $20,250 (d) $23,500 (e) not given

8. The number of minutes spend on the phone per day by a San Diego teenager is 40 minutes, 108 minutes, 62 minutes, 12 minutes, 24 minutes, and 31 minutes.
    (a) 24   (b) 31   (c) 35.5   (d) 40   (e) not given

9. 37, 39, 46, 53, 57, 60, 63, 60, 63, 60.
    (a) 53   (b) 55   (c) 57   (d) 58.5   (e) 60

10. 1400, 1329, 1200, 1386, 1427, 1350.
    (a) 1368   (b) 1350   (c) 1329   (d) 1360   (e) not given

11. 0.25, 0.12, 0.35, 0.43, 0.28.
    (a) 0.26   **(b)** 0.28   **(c)** 0.35   **(d)** 0.43   **(e)** not given

12. The costs of six cars recently purchased by the Weston Company were $18,270, $11,300, $16,400, $9,100, $12,450, and $13,800.
    (a) $1250   **(b)** $13,800   **(c)** $13,125   **(d)** $13,800   **(e)** not given

13. The ages of 10 people swimming laps at the YMCA pool one morning: 60, 18, 24, 36, 39, 32, 70, 12, 15, and 85.
    (a) 21   **(b)** 32   **(c)** 36   **(d)** 37.5   **(e)** not given

For 14 – 19, find the mode:

14. 60, 65, 68, 60, 72, 59, 80.
    (a) 60   **(b)** 65   **(c)** 68   **(d)** 70   **(e)** not given

15. 121, 150, 116, 150, 121, 181, 117, 123.
    (a) 121   **(b)** 150   **(c)** both 121 and 150
    (d) 170   **(e)** 135.5

16. The last six bicycles sold at the Skol Bike shop cost $249, $649, $439, $259, $269, and $249.
    (a) 240   **(b)** 249   **(c)** 259   **(d)** 269   **(e)** not given

17. 3, 6, 9, 18, 26.
    (a) 3   **(b)** 6   **(c)** 9   **(d)** no mode   **(e)** not given

18. The last six color television sets sold at the local Circuit City cost $315, $430, $515, $330, $430, and $615.
    (a) 330   **(b)** 430   **(c)** 515   **(d)** 335   **(e)** not given

19. 18.1, 19.2, 16.5, 18, 19.0, 18.2.
    (a) 18.1   **(b)** 19   **(c)** 19.2   **(d)** 16.5   **(e)** not given

20. During a week in June the 6 PM temperature (in Fahrenheit) were 66°, 71°, 61°, 71°, 71°, 73°, and 75°. What was the average 6 PM temperature of that week?
    (a) 79°   **(b)** 84°   **(c)** 83° **(d)** 77°   **(e)** not given

SECTION B. Applied Problems

EXAMPLE 1: Harvey got a 78 on his last math test. There were 10 other students in his class. Their test scores were 96, 66, 89, 91, 85, 90, 75, 56, 40, and 83. Is Harry above average?

SOLUTION: To find the class average (mean), find the sum of 11 test scores (Harry's plus those of his 10 classmates) and divide by 11:

mean: $\dfrac{78+96+66+89+91+85+90+75+56+40+83}{11} \approx 77.18$

Since Harry's 78 is a bit above the average 77.18, we can say that Harry is doing a bit above average. However, if we find the median of all 11 scores:

$$40, 56, 66, 75, 78, \underset{\substack{\text{middle} \\ \text{value}}}{\underline{83}}, 85, 89, 90, 91, 96$$

The median is 83 and here, Harry's score is in the lower half of the class, which some would consider below average.

EXAMPLE 2: If the average of 4 numbers is 6, find the sum of these four numbers.

SOLUTION: Since the mean is the sum of four numbers divided by 4, i.e.,

mean $= \dfrac{\text{sum of the four numbers}}{4}$, then the sum $= 4$ times the mean. Hence, sum $= 4 \times 6 = 24$.

EXAMPLE 3: Sara took 5 exams in her algebra class. She achieved the same test score on each of her first 4 tests. She was determined to do better on the fifth exam. She studied hard and her fifth test score was 90. This score was 10 points higher than her average score for her first 4 exams.

    (a) What was the average for the first four test scores?
    (b) What was the sum of the first four test scores?
    (c) What is the average of Sara's five test scores?

SOLUTION:
    (a) Since 90 was 10 points higher than her average score for her first 4 exams, the average of her first four exams was $90 - 10 = 80$.
    (b) The sum of the four test scores is therefore 4 times the average or $4 \times 80 = 320$.
    (c) The sum of the five test scores = sum of the first four tests scores plus the fifth test score $= 320 + 90 = 420$, and therefore the average of the five test scores is

$\dfrac{\text{sum of five test scores}}{5} = \dfrac{420}{5} = 82$.

Exercises: Applied Problems

1. The average of a set of 7 numbers is 10. The sum of these 7 numbers is
   (a) 49   (b) 70   (c) 100   (d) 10   (e) 17

2. The average of a set of 8 numbers is 9. If one of these 8 numbers is omitted and the average of the remaining 7 numbers is 10, what is the value of the number left out?
   (a) 9   (b) 8   (c) 7   (d) 10   (e) 2

3. Bob wanted to find the average of his 5 test scores. He found the sum of 5 test scores but he accidentally divided by 4 instead of 5, and got an 81 (as his average) as a result. The sum of the his 5 test scores is
   (a) 405   (b) 324   (c) 20   (d) 810   (e) 581

4. Carol computed the average of 6 numbers. She got the sum of these numbers and divided by 4 to yield the incorrect average of 12. The correct average of these 6 numbers is
   (a) 12   (b) 15   (c) 24   (d) 8   (e) 10

5. Barbara took 4 exams in her English class. The average of these 4 test scores is 85. She took a fifth test and scored 90%. What is the average of all 5 tests?
   (a) 80   (b) 85   (c) 86   (d) 92   (e) 87.5

6. Jim took 5 exams in his history course. On the fifth exam, his score was 90, which was 10 points higher than the average score for his first 4 exams. How many more points was Jim's exam average than his average for the first four exams?
   (a) 10   (b) 2   (c) 4   (d) 6   (e) 1

7. Susan had 4 exams in her math class. The score on her fourth exam was equal to the average score for her first 3 exams. The sum for all four exam scores was 360. What was the sum of her first 3 exams?
   (a) 250   (b) 260   (c) 270   (d) 300   (e) 380

230

8. If the average of four numbers is 6 and a fifth number, 6, is included in the group of four numbers, what is the average of the expanded group of five numbers?

    (a) 4   **(b)** 5   **(c)** 6   **(d)** 7   **(e)** 8

9. If the sum of a group of numbers is 90, and the average of these numbers is 9, how many numbers are in the group?

    (a) 10   **(b)** 9   **(c)** 90   **(d)** 20   **(e)** 30

10. What is the average of 8 numbers whose sum is 320?

    (a) 20   **(b)** 30   **(c)** 40   **(d)** 80   **(e)** 160

11. If the number 20 is added to a group of 4 numbers and their sum to be increased to 75, then what is the average of the 5 numbers?

    (a) 80   **(b)** 4   **(c)** 15   **(d)** 18.75   **(e)** 11

12. A number is added to a group of 5 numbers. As a result the average of the group increases from 10 to 15. What was the added number?

    (a) 5   **(b)** 15   **(c)** 30   **(d)** 40   **(e)** 45

13. If 25 is added to a group of 5 numbers whose average is 7, what is the average of the group as a result of the addition of the new number 25?

    (a) 35   **(b)** 10   **(c)** 8   **(d)** 9   **(e)** 20

14. How much should be added to a group of 5 numbers to increase the average from 8 to 10?

    (a) 40   **(b)** 50   **(c)** 10   **(d)** 8   **(e)** 20

15. A group of 6 numbers has an average of 15. What should be added to the group so that the resulting group has an average of 20?
  (a) 30   **(b)** 40   **(c)** 50   **(d)** 60   **(e)** 20

16. Susan had an 80 average for her 6 math exams. She added up all her 6 exam scores. What was the sum?
  (a) 180   **(b)** 280   **(c)** 360   **(d)** 480   **(e)** 600

17. What 15 is added to a group of 8 numbers, the average of the group becomes 20. What was the sum of the original 8 numbers?
  (a) 150   **(b)** 160   **(c)** 165   **(d)** 170   **(e)** 180

18. Jason's average for 5 exams in his math class was 72. The teacher said he will drop Jason's lowest score, which was a 40. What is Jason's new average?
  (a) 60   **(b)** 70   **(c)** 80   **(d)** 90   **(e)** 76

19. If in exercise 18, Jason's original average was 80, what would his new average be after his teacher drops his lowest score of 40?
  (a) 60   **(b)** 70   **(c)** 80   **(d)** 90   **(e)** 76

20. Gabriel took 6 exams in a history course. The average was 80. On the last exam he asked the teacher to reconsider an answer to one of the questions. After rereading the exam, the teacher game Gabriel an additional 12 points for that exam. What is Gabriel's average now?
  (a) 78   **(b)** 80   **(c)** 82   **(d)** 84   **(e)** 86

SECTION C. Finding a missing value in a group of values with known average

Sometimes you may want to find a missing item or the value of a number that is unknown among a group of items or numbers where the average of the overall group, including the missing item is known.

EXAMPLE 1: A waiter's average daily tips are $25 for 5 days of work. On Monday, he received $24.40; Tuesday, $25.20; Wednesday, $24.75; Thursday, $25.50. What does the waiter expect to collect on the fifth day if his expectation of $25 average for 5 days is to be maintained?

SOLUTION: If the average of 5 numbers is 25, the sum of the 5 numbers is $5(25) = 125$. The sum of the 4 given numbers are $24.40 + 25.20 + 24.75 + 25.50 = 99.85$. Therefore, the fifth number must be $125 - 99.85 = 25.15$.

EXAMPLE 2: For the month of April, Sue has budgeted $170 per week for her family's food bill. During the first 3 weeks of that month, she spent $188, $184, and $185. What will be her limit for the next week, in order to stay within her budget?

SOLUTION: Average for all four weeks is 170, so that the total amount budgeted for food for 4 weeks is $(170)(4) = \$680$. Sue spent in the first 3 weeks: $\$188 + \$184 + \$185 = \$557$. Thus, her limit in the fourth week is $\$680 - \$557 = \$123$.

From the above two examples, we can formulate a procedure for finding the missing item in a group of items to maintain a given average as follows:
1. Multiply the given average by the number of items in the entire group
2. Find the sum of the known items in the group
3. Subtract the sum found in step 2 from the product found in step 1.

Exercises: Finding a missing value in a group of values with known average

1. Gail has obtained test scores of 96, 97, 90 and 85. What grade must she receive on her next test to earn an average of 91?
     (a) 91   **(b)** 92   **(c)** 94   **(d)** 87   **(e)** 90

2. Tom got 93, 95, 85, and 84. He has one more exam this semester. What grade must Tom get on the last exam to achieve an average of 90 for the semester?
     (a) 89   **(b)** 90   **(c)** 92   **(d)** 94   **(e)** not given

3. A student received the following grades: Composition 101, 85%; Prealgebra, 83%; English, 95%; and Chemistry 01, 74%. What grade must be earned in History to attain an overall average of 87%?
     (a) 85   **(b)** 83   **(c)** 95   **(d)** 74   **(e)** 98

4. Jim took 4 exams. He remembered only 3 scores: 93, 82, and 70. He forgot the score he got on his fourth exam, but he did remember that the average for all 4 exams was 81. What was Jim's score for his fourth exam?
     (a) 70   **(b)** 71   **(c)** 73   **(d)** 77   **(e)** 79

For, 5 – 8, Carol got 82, 92, and 78 on her 3 math exams so far. She will be taking one more math exam this semester. What is the minimum score she must get on the last exam if she wants an average for all 4 exams to be at least

5. 70
     (a) 28   **(b)** 38   **(c)** 48   **(d)** 70   **(e)** 75

6. 80
     (a) 55   **(b)** 65   **(c)** 68   **(d)** 78   **(e)** 79

7. 85
   (a) 58   **(b)** 68   **(c)** 78   **(d)** 88   **(e)** 92

8. 88
   (a) 70   **(b)** 80   **(c)** 90   **(d)** 95   **(e)** 100

9. During the first four months of the year, Carl averaged 1000 miles a month in his new car. In January, he drove 1200 miles. In February, he drove 900 miles. In April he drove 500 miles. How many miles did Carl travel in the month of March?
   (a) 1000   **(b)** 700   **(c)** 900   **(d)** 1400   **(e)** 1200

10. The average spent by Gail was $500 per month for the past 5 months, April, May, June, July, and August. In 3 months, June, July, and August, she spent a total of $1700. How much did she spend for the two months of April and May combined?
   (a) $600   **(b)** $700   **(c)** $800   **(d)** $900   **(e)** $1000

For 11 – 15, Mick's average yearly salary for the past five years, 2001, 2002, 2003, 2004, 2005, is $70,000. Find his average salary for the past two years, 2004 and 2005, if the sum of his salaries for the first 3 years, 2001, 2002, and 2003, was

11. $150,000
   (a) $70,000   **(b)** $80,000   **(c)** $90,000   **(d)** $100,000   **(e)** $110,000

12. $180,000
   (a) $75,000   **(b)** $85,000   **(c)** $95,000   **(d)** $105,000   **(e)** $115,000

13. $200,000
   (a) $75,000   **(b)** $85,000   **(c)** $95,000   **(d)** $105,000   **(e)** $115,000

14. $250,000
   (a) $40,000   **(b)** $50,000   **(c)** $60,000   **(d)** $70,000   **(e)** $75,000

15. $300,000
   (a) $55,000   **(b)** $45,000   **(c)** $35,000   **(d)** $25,000   **(e)** $20,000

For 16 – 20, the average weight of 20 students in the chemistry class is 130 pounds. How much does Jeff, the smallest member of the class, weight if the total weight of all the other 19 students is

16. 2450 pounds (lbs)
   (a) 80 lbs   **(b)** 100 lbs   **(c)** 110 lbs   **(d)** 120 lbs   **(e)** 150 lbs

17. 2460 pounds
   (a) 90 lbs   **(b)** 110 lbs   **(c)** 120 lbs   **(d)** 130 lbs   **(e)** 140 lbs

18. 2470 pounds
   (a) 90 lbs   **(b)** 110 lbs   **(c)** 120 lbs   **(d)** 130 lbs   **(e)** 140 lbs

19. 2480 pounds
   (a) 90 lbs   **(b)** 110 lbs   **(c)** 120 lbs   **(d)** 130 lbs   **(e)** 140 lbs

20. 2500 pounds
   (a) 90 lbs   **(b)** 100 lbs   **(c)** 120 lbs   **(d)** 130 lbs   **(e)** 140 lbs

SECTION D. Combining Averages

If we know the average of one group of items and the average of another group and we want to find the average of the combined groups, we could be asked to combine averages. The next example outlines a procedure to accomplish this.

EXAMPLE 1: One class of 25 students took a math test. The average score was 80. Another class of 20 students then took the same math test. The average for this second group was 85. What is the average score of all 45 students?

SOLUTION: The sum of the 25 scores whose average was 80 is $(80)25 = 2000$. The sum of the 20 scores whose average was 85 is $85(20) = 1700$. The sum of all 45 scores is therefore $2000 + 1700 = 3700$ and thus the average $= \dfrac{3700}{45} \approx 82.22$.

From the above example, we can formulate a procedure for finding the average of the combined groups from information given about each individual group as follows:

1. Find the total (in our example above, the total of all grades) by multiplying the average in each group (80 and 85 in our example) by the number of items in each group (25 and 20 in our example).
2. Add the products found in step 1.
3. Divide the total sum found in step 2 by the total number of items in the combined group ($20 + 25 = 45$ in our example).

The basic idea when determining averages is to find the sum of all items we are dealing with and divide by the total number of items that contributed to that total sum. For example, to find the average speed, we must realize that when we say a car traveled 100 miles in 5 hours, we do not know how fast the car was traveling each minute throughout the 5 hours. It could have traveled very rapidly the first hour and very slowly the second hour. All we know is that it took a total of 5 hours to go 100 miles. If we assume that the car traveled at the same speed the entire 5 hours, we then say that the car traveled at the constant speed of $\dfrac{100}{5} = 20$ miles each hour or 20 <u>miles</u> per <u>hour</u> (20 mph). This 20 miles per hour is called the average speed of the car for those 5 hours. We can think of it in the following way: 100 is the sum of 5 numbers, the first number is the number of miles traveled in the first hour, the second number is the number of miles traveled in the second hour and so on. The sum of these five numbers is 100. The average of these five numbers is: (the sum of the 5 numbers) $\div 5$.

In general, the average speed $= \dfrac{\text{total distance traveled}}{\text{total number of hours required to travel the total distance}}$.

Similarly, the average score $= \dfrac{\text{total sum of all scores}}{\text{total number of scores contributing to the total}}$,

237

and the average cost $= \dfrac{\text{total cost of all items}}{\text{total number of items contributing to the total cost}}$.

Conversely, if we know the average and the number of items, we can get the total distance traveled, total cost of all the items, etc., by multiplying the average by the total number of items contributing to the total.

EXAMPLE 2: Jason drove for 5 hours at an average speed of 36 miles per hour (mph). He then drove for 10 hours averaging 45 mph. Find his average speed in mph for his 15 hour trip.

SOLUTION: Jason traveled a total of $5 + 10 = 15$ hours. During the first part of his trip, Jason covered a distance of $5(36) = 180 \text{ miles}$. During the second part of his trip, he traveled $10(45) = 450 \text{ miles}$. Hence, the distance traveled for the entire trip is $180 + 450 = 630 \text{ miles}$. The average speed is $\dfrac{\text{the total distance traveled}}{\text{total time spent traveling}} = \dfrac{630}{15} = 42 \text{ mph}$.

EXAMPLE 3: Rina bought a large mixture of candy consisting of three candy types A, B, and C. In this mixture there are: 50 pounds of type A costing 75¢ per pound, 40 pounds of type B costing 60¢ per pound, and 60 pounds of type C costing \$1.10 per pound. Find the average cost per pound of candy.

SOLUTION:
$$\begin{aligned}
\text{Total cost} &= \text{cost of type A candy} + \text{cost of type B candy} + \text{cost of type C candy} \\
&= (50 \text{ pounds} \times 75¢) + (40 \text{ pounds} \times 60¢) + (60 \text{ pounds} \times \$1.10) \\
&= 37.50 + 24.00 + 66.00 \\
&= \$127.50
\end{aligned}$$
The total number of pounds of candy $= 50 + 40 + 60 = 150$.
The average cost is therefore $\dfrac{\$127.50}{150} = \$0.85$

EXAMPLE 4: A car travels an average 40 mph for 7 hours. How many miles did it travel?

SOLUTION: Average $= \dfrac{\text{total miles traveled}}{\text{total time}}$ and so total miles $= \text{average} \times \text{total time}$.
Total miles $= 40 \times 7 = 280 \text{ miles}$.

EXAMPLE 5: Gabriel traveled 150 miles at an average speed of 50 mph. How long did it take him?

SOLUTION: Average speed $= \dfrac{\text{total distance}}{\text{total time taken to travel distance}}$. Therefore, $50 = \dfrac{150}{\text{total time}}$

and total time $\times 50 = 150$, so that total time $= \dfrac{150}{50} = 3$ hours .

EXAMPLE 6: A car traveled at 50 mph for 50 miles, then 100 miles at 25 mph, then 30 miles at 30 mph. What is the average rate of speed (in mph) for the entire trip?

SOLUTION: Total distance: $50 + 100 + 30 = 180$ miles .

Total time for trip: $\dfrac{50 \text{ miles}}{50 \text{ mph}} = 1$ hour ; $\dfrac{100 \text{ miles}}{25 \text{ mph}} = 4$ hours ; $\dfrac{30 \text{ miles}}{30 \text{ mph}} = 1$ hour .

So that total time $= 1 + 4 + 1 = 6$ hours .

Therefore the average speed $= \dfrac{\text{total distance}}{\text{total time}} = \dfrac{180 \text{ miles}}{6 \text{ hours}} = 30$ mph .

Exercises: Combining Averages

1. In a chemistry class of forty students, the average test score for twenty-five girls is 75, and for the fifteen boys is 67. Find the average score for the entire class.
    (a) 75   **(b)** 67   **(c)** 71   **(d)** 72   **(e)** 83

2. In a college history class, the average age of eight girls is 17 years, and the average age of twelve boys is 22 years. Which of the following computation represents the average age, in years, for the entire class, i.e. for all twenty students in the class?
    (a) $\dfrac{17+22}{2}$ **(b)** $\dfrac{17+22}{8+12}$ **(c)** $\dfrac{(8\times17)+(12\times22)}{8+12}$
    (d) $\dfrac{(12\times17)+(8\times22)}{8+12}$ **(e)** $\dfrac{(8\times17)+(12\times22)}{8\times12}$

3. Michael takes a long trip with his wife, Karen, and their five kids. He travels 100 miles at 40 mph. The next 200 miles, he speeds up and travels at 50 mph. For the next 300 miles, he slows down and travels at 30 mph. How long did the entire trip of 600 miles take Michael?
    (a) 2.5 hours **(b)** 6 hours   **(c)** 10 hours   **(d)** 16.5 hours **(e)** 32.4 hours

4. On his way back, Michael travels 480 miles at 60 mph and 120 miles at 30 mph. Approximately what was his average speed, in mph, for the entire round trip?
    (a) 40 mph   **(b)** 45 mph   **(c)** 30 mph   **(d)** 42 mph   **(e)** 50 mph

5. A total of 100 freshmen and juniors were given an IQ test. The 60 freshmen attained an average score of 105, while the 40 juniors attained an average of 110. What was the average IQ score for all 100 students who took the IQ test?
    (a) 107.5   **(b)** 105   **(c)** 110   **(d)** 205   **(e)** 107

6. During a trip that lasted seven hours, Daniel averaged 34 mph for the first three hours, his wife, Jennifer, drove the remainder of the trip at 62 mph. Find the average speed in mph for the entire trip.
    (a) 34   **(b)** 62   **(c)** 48   **(d)** 50   **(e)** not given

7. On a particular turnpike, the toll is 5¢ a mile for the first 40 miles, and 2¢ a mile for the remaining 160 miles on a 200 mile trip. Find the average cost, in cents per mile, for the entire trip.
   (a) 2.6¢ per mile　　　　　(b) 3.5¢ per mile　　　　　(c) 5¢ per mile
   (d) 2¢ per mile　　　(e) not given

8. Gail, a brilliant, gorgeous resident of radiology at a very prestigious hospital in New York, made a three hour trip at an average speed of 60 mph. The first hour, she averaged 70 mph, the next hour and a half she averaged 50 mph. Find her average speed for the remaining part of her trip.
   (a) 70 mph　　(b) 50 mph　　(c) 35 mph　　(d) 60 mph　　(e) not given

9. A plane travels from city A to city B. It leaves city A at 1 PM and traveled 455 miles and arrived at city B at 4:30 PM that same day. The plane's average speed, in mph, was
   (a) 152 mph　　(b) 90 mph　　(c) 130 mph　　(d) 455 mph　　(e) not given

10. A school basketball team played on five different occasions with a total attendance of 2640 paid spectators for all five games. What is the average attendance for each game?
    (a) 264　　　(b) 528　　　(c) 500　　　(d) 275　　　(e) not given

11. On four shopping days, Rina purchased two dozen oranges at 93¢ per dozen on Monday; on Tuesday, one dozen oranges at 96¢ per dozen; on Wednesday one dozen oranges at 94¢ per dozen; and on Thursday, she purchased three dozen at $1.01 per dozen. The average cost per dozen oranges is:
    (a) 97¢　　　(b) 95.5¢　　(c) 90¢　　(d) 95.75¢　　(e) 94¢

12. While on a trip, Jack made the following purchases of gasoline: eight gallons at $1.96 per gallon, twelve gallons at $2.06 per gallon, and five gallons at $1.92 per gallon. Jack's average cost per gallon was approximately
    (a) $2.00　　(b) $1.98　　(c) $2.02　　(d) $1.95　　(e) $1.99

13. Jane took thirteen quizzes. She received three 70's, three 75's, four 80's, a 91, and three 95's. Her average quiz score was:
    (a) 87　　　(b) 82.2　　(c) 80　　(d) 87.5　　(e) 92

14. On eight math quizzes, Carol averaged a score of 80. However, for the next four quizzes she scored a 50, 70, 75, and 65. Find her average for all 12 quizzes.
   (a) 72.5   **(b)** 65   **(c)** 80   **(d)** 75   **(e)** not given

15. The Red Chicken Poultry farm sold 100 dozen eggs at 90¢ a dozen, 80 dozen at 85¢ a dozen, 90 dozen at $1.00 a dozen. The average selling price per dozen is approximately:
   (a) 92¢   **(b)** 90¢   **(c)** 94¢   **(d)** 96¢   **(e)** 98¢

16. Bob received the following grade on six exams in a math course: 61, 73, 80, 66, 70, and 94. To complete his final letter grade in the course, the instructor averages the grades on the six exams and assigns a letter grade as follows:
   A, if the average is greater than or equal to 90 and less than or equal to 100,
   B, if the average is greater than or equal to 80 and less than or equal to 89,
   C, if the average is greater than or equal to 70 and less than or equal to 79,
   D, if the average is greater than or equal to 60 and less than or equal to 69,
   F, if the average is 59 or less.
   What is Bob's letter grade for the course?
   (a) A   **(b)** B   **(c)** C   **(d)** D   **(e)** F

17. A class of 34 students took a history exam. Six students scored 96%, ten students scored 80%, four scored 70%, six scored 60%, and eight scored 54%. The class average is:
   (a) 72   **(b)** 66   **(c)** 75   **(d)** 80   **(e)** 70

18. In some western college, at a particular time, there are six English classes with an average of 18 students each, nine math classes with an average of 20 students each, ten history classes with an average of 30 students each, and twelve chemistry classes with an average of 25 students each. The total number of students enrolled in these 37 classes is:
   (a) 93   **(b)** 888   **(c)** 370   **(d)** 24   **(e)** 908

19. Referring to exercise 18, what is the average class size, i.e. what is the average number of students in each class?
   (a) 18   **(b)** 25   **(c)** 26   **(d)** 24   **(e)** 23.25

20. James bought five items at store A at an average price of $6 per item and then went to shop in store B and bought seven items at an average price of $18 per item. What was the average price James paid for the total of 12 items he bought?
   (a) $12   **(b)** $11   **(c)** $13   **(d)** $10   **(e)** $14

Solutions to Exercises

Chapter 1 – Integers

A. Whole Numbers

1. 4 100's; 5 10's; 3 ones

3. 1 10,000; 2 1000's; 3 100's; 4 10's; 5 ones

5. 300 + 40 + 5

7. 10,000 + 3000 + 400 + 50 + 6

9. Five hundred and seventy-six

11. Five hundred sixty-seven thousand three hundred forty-five

13. 1156

15. 7536

17. 491

19. 9870

21. 56,000

23. 56,000

25. 1368

27. 10,396

29. 21,394

31. 789 + (105 + 195) = 789 + 300 = 1089

33. 60 + 100

35. 157 R 15

37. 18 R 6

39. 34 R 5

41. 30

43. c

45. d

47. e

49. d

51. d

53. e

55. b

57. a

B and C. Introducing Integers and Absolute Values

1. $-3°$

3. A loss of $5

5. $-9°$

7. $7 > -5$

9. $4 > -10$

11. $-5, -3, -2, 0, 3, 6, 8$

13. 5

15. -6

17. 3

19. 12

21. $|-1| > -|5|$

23. 6

D. Operations with Integers

1. 5	**3.** 9	**5.** −5	**7.** −53
9. −171	**11.** 19	**13.** 26	**15.** 12
17. 48	**19.** −117	**21.** 96	**23.** −17
25. −5	**27.** −51	**29.** −$24	**31.** e
33. c	**35.** e	**37.** d	

E. Order of Arithmetic Operations and Exponents

1. 19	**3.** 17	**5.** −10	**7.** 8
9. $(-2)^4$	**11.** $2 \cdot 2 \cdot 2$	**13.** $3 \cdot 3 \cdot 7$	**15.** 28
17. −9	**19.** 8	**21.** −48	**23.** −16

F. Multiples and Factors of Integers

1. 1, 5	**3.** 1, 2, 3, 4, 6, 12	**5.** An integer greater than one, whose only divisors are one and itself.	
7. $3 \cdot 5$	**9.** $2 \cdot 3 \cdot 7$	**11.** 5	**13.** 15
15. 25	**17.** 21	**19.** 42	**21.** 90
23. e	**25.** d	**27.** e	

G. Translating a Statement into a Mathematical Expression

1. $n+5$	**3.** $5n$	**5.** $m+x$	**7.** $\dfrac{x}{2}$
9. $p+10$	**11.** $7-y$	**13.** $15+3x$	**15.** $7y+x$
17. $10+12x$	**19.** e	**21.** c	**23.** 5
25. −30			

H. Solving Applied Problems

1.	a	**3.**	b	**5.**	b	**7.**	a
9.	a	**11.**	d	**13.**	b	**15.**	e
17.	b	**19.**	d	**21.**	b	**23.**	c

Chapter 2 – Working with Fractions and Mixed Numbers

B. Expressing an Improper Fraction as a Mixed Number and a Mixed Number as an Improper Fraction

1. 0   3. 1   5. undefined   7. b

9. d   11. a   13. d   15. a

17. b   19. b   21. c

C. Equivalent Fractions; Reducing Fractions to Lowest Terms

1. $\dfrac{21}{49}$ 3. $\dfrac{15}{20}$ 5. $\dfrac{27}{39}$ 7. $\dfrac{70}{80}$

9. $\dfrac{3}{5}$ 11. $\dfrac{3}{4}$ 13. $\dfrac{5}{6}$ 15. $\dfrac{6}{7}$

17. $\dfrac{7}{5}$ or $1\dfrac{2}{5}$ 19. $\dfrac{5}{4}$ or $1\dfrac{1}{4}$ 21. $-\dfrac{2}{3}$ 23. $-\dfrac{7}{8}$

D. Multiplying and Dividing Fractions and Mixed Numbers

1. $\dfrac{1}{15}$ 3. $\dfrac{5}{24}$ 5. $\dfrac{1}{7}$ 7. 2

9. $-\dfrac{8}{11}$ 11. $\dfrac{10}{7}$ 13. $\dfrac{2}{81}$ 15. 5

17. $\dfrac{41}{4}$ or $10\dfrac{1}{4}$ 19. 6   21. $\dfrac{215}{6}$ or $35\dfrac{5}{6}$ 23. $\dfrac{4}{5}$

25. $\dfrac{9}{16}$ 27. $-\dfrac{1}{10}$ 29. $\dfrac{9}{16}$ 31. 27

33. $\dfrac{1}{52}$ 35. 6   37. $\dfrac{1}{24}$ 39. $\dfrac{25}{144}$

E. Solving Application Problems

1. a	**3.** d	**5.** d	**7.** b
9. e	**11.** d	**13.** a	**15.** c
17. a	**19.** a	**21.** c	**23.** b
25. c			

F. Adding and Subtracting Fractions and Mixed Numbers

1. $\dfrac{2}{45}$ **3.** $\dfrac{1}{2}$ **5.** $\dfrac{53}{56}$ **7.** $\dfrac{49}{30}$ or $1\dfrac{19}{30}$

9. $-\dfrac{11}{20}$ **11.** $\dfrac{3}{26}$ **13.** $-\dfrac{15}{28}$ **15.** $-\dfrac{21}{80}$

17. $\dfrac{57}{100}$ **19.** $\dfrac{31}{32}$ **21.** $18\dfrac{11}{12}$ **23.** $5\dfrac{7}{10}$

25. $3\dfrac{1}{2}$ **27.** $1\dfrac{13}{24}$ **29.** $19\dfrac{11}{12}$ **31.** $21\dfrac{1}{9}$

33. a   **35.** d   **37.** c   **39.** b

G. Solving Application Problems

1. c	**3.** a	**5.** d	**7.** b
9. c	**11.** b	**13.** a	**15.** a
17. b	**19.** b	**21.** a	

H. Comparing Fractions, Complex Fractions and Order of Operation

1. $\dfrac{19}{40}$ **3.** $\dfrac{13}{16}$ **5.** $\dfrac{7}{15}$ **7.** $\dfrac{19}{24}$

9. $\dfrac{9}{14}$ **11.** b   **13.** c   **15.** a

17. c   **19.** e

Chapter 3 – Working with Decimals

A. Understanding Decimals and Decimal Fractions

1. 0.67   **3.** 45.6   **5.** 984.5   **7.** 0.567

9. $\dfrac{23}{100}$ **11.** $\dfrac{5}{10,000}$ **13.** $8\dfrac{9}{100}$ **15.** $\dfrac{67}{1000}$

17. Nine tenths   **19.** Five thousandths

21. Eight tenths   **23.** Three ten-thousandths

25. $5(1)+3\left(\dfrac{1}{10}\right)+6\left(\dfrac{1}{100}\right)$ **27.** $5\left(\dfrac{1}{10}\right)+6\left(\dfrac{1}{100}\right)$

29. $7\left(\dfrac{1}{100}\right)+8\left(\dfrac{1}{1000}\right)$ **31.** $2\left(\dfrac{1}{10}\right)+3\left(\dfrac{1}{100}\right)+4\left(\dfrac{1}{1000}\right)+5\left(\dfrac{1}{10,000}\right)$

33. $\dfrac{5}{10}$ **35.** $\dfrac{5}{1000}$ **37.** $\dfrac{1}{10}$ **39.** $\dfrac{1}{1000}$

B and C. Ordering and Comparing Decimals and Rounding

1. <   **3.** >   **5.** <

7. 0.00999, 0.099, 0.109, 0.110, 0.19   **9.** No. $4.10 = 4\dfrac{10}{100} = 4\dfrac{1}{10} = 4.1$

11. No   **13.** 0.10, 0.12, 0.13   **15.** 0.99, 0.98

17. .2, .20, 0.20, $\dfrac{2}{10}$, $\dfrac{20}{100}$, two tenths, 20 hundredths, 200 thousandths

19. 4   **21.** 3.23   **23.** 6.8   **25.** 4.10

27. 4   **29.** a

D. Adding and Subtracting Decimals

1. 24.523	**3.** 13.601	**5.** 5.3	**7.** 0.9
9. 0.94	**11.** 0.99	**13.** 7.98	**15.** 34.995
17. 81.662	**19.** 85.267	**21.** 27.7	**23.** 0.12
25. c	**27.** d	**29.** e	

E. Multiplying and Dividing Decimals

1. 8	**3.** 0.00000012	**5.** 0.0075	**7.** 0.00105
9. 0.54	**11.** 21,640.63	**13.** 296.571	**15.** 456.7
17. 6580	**19.** 2.34	**21.** 30,700	**23.** 0.4
25. 0.9	**27.** 500	**29.** 0.86	**31.** 0.045
33. 54	**35.** 7.5	**37.** 2.8123	**39.** 0.3562
41. 65.96	**43.** 0.14	**45.** 12.246	**47.** d
49. d	**51.** c	**53.** b	**55.** b
57. d	**59.** 1.4	**61.** 1.5	

F. Converting Fractions to Decimals; Converting Decimals to Fractions

1. 0.25	**3.** 0.2	**5.** $\frac{2}{5}$	**7.** $\frac{3}{5}$
9. $\frac{3}{20}$	**11.** $2\frac{1}{4}$	**13.** 0.375	**15.** 0.55
17. 0.175	**19.** 1.4	**21.** $0.\overline{636}$	**23.** $0.8\overline{3}$
25. $0.\overline{15}$	**27.** $5.58\overline{3}$	**29.** 0.214	**31.** 0.882
33. 0.130	**35.** 0.429		

37. Yes. $\frac{5}{8} = 0.625$, $0.625 - 0.6 = 0.025$. The drill bit is 0.025 in larger than the cable.

A–7

G. Solving Application Problems

1.	d	**3.**	c	**5.**	a	**7.**	d
9.	e	**11.**	e	**13.**	e	**15.**	b
17.	a	**19.**	b	**21.**	d	**23.**	a
25.	a	**27.**	b	**29.**	e	**31.**	a
33.	d						

Chapter 4 – Exponents, Scientific Notation, and Square Roots

A. Positive Powers

1. a	**3.** a	**5.** b	**7.** b
9. a	**11.** a	**13.** c	**15.** d
17. b	**19.** c		

B. Negative Powers

1. d	**3.** d	**5.** d	**7.** a
9. b	**11.** a	**13.** a	**15.** c
17. d	**19.** b	**21.** c	**23.** d
25. a			

C. Arithmetic Operations Using Scientific Notation

1. b	**3.** a	**5.** b	**7.** a
9. b	**11.** a	**13.** a	**15.** e
17. a	**19.** b		

E. Approximating Square Roots

1. 3, –3	**3.** 7, –7	**5.** 6	**7.** 9
9. –7	**11.** 0.9	**13.** 30	**15.** –100
17. 12	**19.** 0.07	**21.** 130	**23.** c
25. 6	**27.** Does not exist	**29.** Does not exist	**31.** Does not exist
33. c	**35.** b	**37.** c	**39.** b
41. b	**43.** b		

F. Simplifying Square Roots

1. b	**3.** d	**5.** b	**7.** c
9. b	**11.** b	**13.** a	**15.** a
17. b	**19.** b		

G. Adding and Subtracting Like Square Roots

1. c	**3.** a	**5.** e	**7.** a
9. a	**11.** b	**13.** c	**15.** a
17. c	**19.** d		

Chapter 5 – Ratios and Proportions

A. Ratios

1. a	**3.** d	**5.** b	**7.** c
9. e	**11.** a	**13.** e	**15.** e
17. c	**19.** b		

B. Rates

1. $21\frac{11}{19}$ calories per gram of fat	**3.** $22\frac{1}{2}$ mi/gal	**5.** \$8 per hour	
7. $53\frac{1}{3}$ mph	**9.** e	**11.** e	**13.** c
15. c	**17.** c	**19.** c	**21.** c
23. d	**25.** b	**27.** b	**29.** a
31. a			

C. Proportions

1. True	**3.** True	**5.** True	**7.** Not true
9. True	**11.** True	**13.** True	**15.** Not true
17. True	**19.** True		

D. Writing and Reading a Proportion

1. $\dfrac{4}{9}=\dfrac{28}{63}$

3. $\dfrac{\frac{1}{3}}{\frac{1}{8}}=\dfrac{\frac{1}{4}}{\frac{3}{32}}$

5. $\dfrac{3}{8}=\dfrac{18}{48}$

7. $\dfrac{12}{7}=\dfrac{48}{28}$

9. $\dfrac{7}{11}=\dfrac{8}{12}$

11. $\dfrac{2\text{ printers}}{4\text{ secretaries}}=\dfrac{12\text{ printers}}{24\text{ secretaries}}$

13. $\dfrac{\$6}{4\text{ pounds}}=\dfrac{\$264}{176\text{ pounds}}$

15. $\dfrac{4\text{ baskets}}{7\text{ free throws}}=\dfrac{12\text{ baskets}}{21\text{ free throws}}$

17. $\dfrac{6\text{ benches}}{24\text{ people}}=\dfrac{48\text{ benches}}{192\text{ people}}$

19. $\dfrac{176\text{ miles}}{8\text{ gallons}}=\dfrac{528\text{ miles}}{24\text{ gallons}}$

E. Solving $ax = b$

1. $x=13$ **3.** $y=-5$ **5.** $x=13$ **7.** $x=13$

9. $x=-3$ **11.** $y=15$ **13.** $x=5$ **15.** $a=-7$

17. $y=-11$ **19.** $x=11$ **21.** $y=-4$ **23.** $x=-12$

F. Solving a Proportion

1. $x=3$ **3.** $x=20$ **5.** $x=8$ **7.** $n=5\frac{1}{3}$

9. $n=14$ **11.** $n=8$ **13.** $y=2$ **15.** b

17. b   **19.** b

G. Solving Applied Problems Involving Proportions

1. c   **3.** c   **5.** e   **7.** c

9. d   **11.** e   **13.** c   **15.** b

17. b   **19.** c   **21.** c   **23.** d

25. c   **27.** b   **29.** d   **31.** c

33. c   **35.** d

Chapter 6 – Percentages

A. Understanding the Meaning of a Percent
1. 31% of the students voted
3. 63% of the power boats had a radar navigation system
5. 16% of people use electric toothbrushes
7. 11% are defective
9. 160%
11. 31% did not vote
13. 58% is water
15. $14\frac{1}{2}\%$
17. 84%
19. a

B. Converting a Decimal to Percent and Percent to Decimal
1. 0.538
3. 0.0233
5. 0.244
7. 0.052
9. 47.6%
11. 13.12
13. 5%
15. 52%
17. 0.05
19. 1.05

D. Changing Fractions, Decimals, and Percents
1. d
3. c
5. c
7. a
9. c
11. b
13. a
15. a
17. d
19. a

E. Setting Up and Solving Percent Problems
1. a
3. b
5. b
7. c
9. b
11. a
13. b
15. c
17. a
19. a
21. b

F. Solving Applied Problems Involving Percent

1. c	3. d	5. d	7. c
9. c	11. a	13. b	15. b
17. b	19. b	21. c	23. a
25. c	27. a	29. c	31. c
33. e	35. a	37. c	39. d
41. c	43. a	45. b	47. b
49. e			

Chapter 7 – Averages

A. Mean, Median, Mode
1.	c	**3.**	b	**5.**	e	**7.**	c
9.	d	**11.**	b	**13.**	e	**15.**	a
17.	d	**19.**	e				

B. Applied Problems
1.	b	**3.**	b	**5.**	c	**7.**	c
9.	a	**11.**	c	**13.**	b	**15.**	c
17.	c	**19.**	d				

C. Finding a Missing Value in a Group of Values with Known Average
1.	d	**3.**	e	**5.**	a	**7.**	d
9.	d	**11.**	d	**13.**	a	**15.**	d
17.	e	**19.**	c				

D. Combining Averages
1.	d	**3.**	d	**5.**	e	**7.**	a
9.	c	**11.**	a	**13.**	a	**15.**	a
17.	a	**19.**	d				